Stefan Schmaus

Spintronics with individual metal-organic molecules

Experimental Condensed Matter Physics
Band 2

Herausgeber
Physikalisches Institut
Prof. Dr. Hilbert von Löhneysen
Prof. Dr. Alexey Ustinov
Prof. Dr. Georg Weiß
Prof. Dr. Wulf Wulfhekel

Spintronics with individual metal-organic molecules

by
Stefan Schmaus

Dissertation, Karlsruher Institut für Technologie
Fakultät für Physik
Tag der mündlichen Prüfung: 17.12.2010
Referenten: Prof. Dr. Wulf Wulfhekel, apl. Prof. Ferdinand Evers

Impressum

Karlsruher Institut für Technologie (KIT)
KIT Scientific Publishing
Straße am Forum 2
D-76131 Karlsruhe
www.ksp.kit.edu

KIT – Universität des Landes Baden-Württemberg und nationales
Forschungszentrum in der Helmholtz-Gemeinschaft

KIT Scientific Publishing 2011
Print on Demand

ISSN 2191-9925
ISBN 978-3-86644-649-6

Spintronics with individual metal-organic molecules

Zur Erlangung des akademischen Grades eines

DOKTORS DER NATURWISSENSCHAFTEN

vor der Fakultät für Physik des
Karlsruher Instituts für Technologie

akzeptierte

DISSERTATION

von

Stefan Schmaus
aus Ehingen (Donau)

Tag der mündlichen Prüfung   17.12.2010

Referent   Prof. Wulf Wulfhekel
Korreferent   apl. Prof. Ferdinand Evers

Contents

Deutsche Zusammenfassung

Die moderne Informationsgesellschaft benötigt immer leistungsfähigere und gleichzeitig auch günstige Bauelemente für integrierte Schaltkreise und Datenspeicher. Dabei gelang es der Halbleiterindustrie bisher, diese Anforderungen durch eine effiziente Optimierung bestehender Technologien zu erreichen. Es lässt sich jedoch abschätzen, dass in naher Zukunft eine einfache Verkleinerung der Bausteine, die im Zentrum aktueller Verbesserungen steht, an ihre Grenzen stoßen wird. In den inzwischen erreichten Größen weniger Nanometer spielen Quanteneffekte eine immer wichtigere Rolle. Die auf der Halbleiterelektronik basierenden Technologien lassen sich in diesen Größen nicht mehr eins-zu-eins umsetzen. Es ist deshalb von entscheidender Bedeutung für die weitere Entwicklung, neue Konzepte einzuführen, die diese Probleme nicht nur lösen sondern sich diese Effekte sogar zu Nutze machen. Es bleibt ein zentraler Punkt der Forschung, diese Ansätze weiter zu entwickeln und eine Verwendung in kommerziellen Anwendungen voranzutreiben.

Ein auf solchen neuen Effekten beruhendes Konzept, das aus der heutigen Informationsgesellschaft nicht mehr wegzudenken ist, sind die Magnetowiderstände. In der modernen Datenspeicherung spielt der Riesenmagnetowiderstand (*giant magnetoresistance:* GMR) seit seiner Entdeckung 1988 eine entscheidende Rolle [1, 2]. Seine Anwendung in Leseköpfen von Festplatten ermöglichte es überhaupt erst, heutige Speicherdichten ($> 375^{Gb}/in^2$ [3]) zu erreichen. Die Anwendung des GMRs in Festplatten ist ein gutes Beispiel für eine schnelle Umsetzung von der Grundlagenforschung zur kommerziellen Anwendung. Das Hauptaugenmerk der intensiven Forschung auf dem Gebiet der Magnetowiderstände lag auf der Verbesserung des Widerstandsverhältnisses, um noch höhere Speicherdichten zu erreichen. Einen weiteren Schub lieferten dabei Systeme, die nicht mehr auf dem Riesen- sondern auf dem Tunnelmagnetowiderstand (*tunneling magnetoresistance:* TMR) beruhen [4, 5]. In den auf Eisen-Magnesiumoxid-Eisen-Kontakten basierenden Komponenten wurde ein deutlich höherer Magnetowiderstand gemessen. Diese verdrängten GMR-basierte Bauelemente und wurden in der Folge in kommerziellen Festplatten verwendet. Ein

Problem der TMR-basierten Komponenten ist die geringe Stromdichte, da der Elektronentransport zwischen den Elektroden mittels Tunneleffekt im Vergleich zum metallischen Transport deutlich reduziert ist. Der Durchlasswiderstand steigt dabei so stark an, dass bei kleinsten Nanostrukturen die Ströme zu niedrig für hochfrequente Anwendungen werden. Deshalb wird in aktuellen Anwendungen wieder verstärkt auf GMR-basierte Komponenten zurückgegriffen [6]. Dabei muss der Fokus nun darauf gelegt werden, Systeme zu entwickeln, welche hinreichend große Magnetowiderstände haben und gleichzeitig klein genug für die gewünschten Anwendungen sind.

Nicht nur in der Datenspeicherung auch in der Elektronik liegt der Fokus auf einer weiteren Verkleinerung der verschiedenen Komponenten. Ein vielversprechender Ansatz dabei, als Nachfolgetechnologie zur Halbleiterelektronik, basiert auf dem theoretischen Konzept von Aviram und Ratner [7]. Sie machten den Vorschlag, einzelne Bauelemente, wie zum Beispiel eine Diode, aus einem einzelnen Molekül herzustellen und diese zu integrierten Schaltkreisen zusammenzufügen. Dieses Konzept leitete eine intensive Forschung auf dem dadurch entstandenen Gebiet der molekularen Elektronik ein. Zu Beginn lag das Hauptproblem darin, geeignete Methoden zur Untersuchung einzelner Moleküle zu entwickeln. In den letzten Jahren haben sich zwei unterschiedliche Ansätze — auf der einen Seite die Bruchkontaktmessungen [8], auf der anderen Kontaktierungen mit dem Rastertunnelmikroskop (*scanning tunneling microscope*: STM) [9] — sowie damit verwandte Methoden durchgesetzt. Darauf basierend gelang es in den letzten Jahren, einige Komponenten aus Einzelmolekülen herzustellen [8, 10–14]. Es bleibt aber weiterhin ein großes Problem die Moleküle in reale Schaltkreise zu integrieren.

Das Ziel dieser Arbeit ist es, diese beiden Konzepte zu verknüpfen und Magnetowiderstandsmessungen an Einzelmolekülen durchzuführen. Dadurch ließen sich GMR-Kontakte in der Größe weniger Nanometer herstellen. Dazu musste eine geeignete Methode entwickelt werden, die es möglich macht, den spin-polarisierten Transport einzelner Moleküle zu messen. Der Ansatz dazu beruht auf der Idee, das spin-polarisierte STM [15] für Kontaktmessungen zu verwenden. Der erste Schritt war, eine geeignete Methode zu etablieren, den Magnetotransport über einzelne Moleküle zu messen. Als Grundlage dazu dienten die Kontaktierungsmessungen mit dem STM von Joachim *et al.* [9]. Diese Methode wurde in den letzten Jahren verfeinert und auf unterschiedliche Moleküle angewendet [16–20]. Die Kontaktmessungen lassen sich dabei sehr gut mit anderen grundlegenden STM-Techniken kombinieren. So lassen sich zum Beispiel Vibrations- oder magnetische Anregungen mit Hilfe der inelastischen Tunnelspektroskopie messen [21–23]. Diese etablierten Techniken sollten in dieser Arbeit mit dem spin-polarisierten STM kombiniert werden, um die ersten GMR-Messungen an Einzelmolekülen durchzuführen.

Nicht nur als mögliche GMR-Bauelemente sind Einzelmoleküle ein vielversprechendes Konzept für die weitere Verbesserung von Speicherdichten in Festplatten. Für Domänengrößen, welche in modernen Festplatten verwendet werden, ist es schwierig eine ausreichend hohe magnetische Stabilität zu erreichen. Der Ansatz, die Stabilität der lokalisierten Spins zu erhöhen, ist, diese durch einen organischen Molekülkäfig zu stabilisieren. Diese Molekülklasse, die einzelmolekularen Magnete, zeigen hohe magnetische Anisotropien welche für stabile Bits essentiell sind. Sollte es gelingen stabile Bits aus den molekularen Magneten herzustellen, ließe sich die Domänengröße auf die Ausdehnung einzelner Moleküle reduzieren. Es ist deshalb von entscheidendem Interesse, die Moleküle auf Oberflächen aufzubringen und zu untersuchen, wie sich dadurch ihre elektronischen und magnetischen Eigenschaften ändern.

Diese Arbeit widmet sich diesen beiden magnetischen Anwendungsmöglichkeiten von Molekülen. In Kapitel 2 werden zuerst die Grundlagen sowie aktuelle Forschungsergebnisse auf dem Gebiet des Magnetismus, der Rastertunnelmikroskopie sowie der molekularen Elektronik präsentiert, welche für die weitere Arbeit von Bedeutung sind. Kapitel 3 zeigt den verwendeten Versuchsaufbau sowie die verschiedenen Materialien welche bei der Durchführung verwendet wurden. Darauf basierend wurden die ersten Transportmessungen an Einzelmolekülen durchgeführt. Phthalocyanine-Moleküle, die aufgrund ihrer Anwendungen in verschiedenen Systemen und Konfigurationen untersucht wurden [24], werden benutzt, um die Transportmesungen zu etablieren. Phthalocyanine bestehen aus vier aromatischen Isoindoldoppelringen (C_8H_4N) die über vier Stickstoffbrücken eine geschlossene aromatische Ringstruktur ausbilden. Durch Platzieren verschiedener Metalle in der zentralen Kavität lassen sich die Eigenschaften der Phthalocyanine verändern. In Kapitel 4 werden die Ergebnisse für verschieden Moleküle (H_2Pc, CoPc und MnPc) auf unterschiedlichen Oberflächen (Cu und Co) gezeigt. Dabei wurden die unterschiedlichen Einflüsse auf den Transport, wie zum Beispiel die Hybridisierung mit den Oberflächen, anhand dieser Systeme untersucht.

Basierend auf diesen Voruntersuchungen wurden magnetische Transportmessungen an H_2Pc-Molekülen durchgeführt (Kapitel 5). Dazu wurde H_2Pc auf magnetische Kobaltnanostrukturen aufgedampft und Transportmessungen mit einer magnetischen Kobaltspitze durchgeführt. So gelang es, die ersten Messungen des GMRs von Einzelmolekülen durchzuführen. Diese bahnbrechenden Ergebnisse könnten einen entscheidenden Beitrag zur fortschreitenden Miniaturisierung von Speichermedien leisten, da sie relativ große Ströme mit Widerstandsverhältnissen verknüpfen konnten, die für moderne Anwendungen benötigt werden.

Theoretische Rechnungen bestätigen die experimentellen Messungen und liefern eine physikalische Erklärung der Beobachtungen. Grund für den starken Einfluss des nichtmagnetischen Moleküls auf den Magnetotransport ist die starke Abhängigkeit der Hybridisierung der Moleküle mit den Elektroden vom Spinkanal. Darüber hinaus wurden GMR-Messungen an asymetrischen $Fe/H_2Pc/Mn$-Konktakten durchgeführt. Aufgrund der unterschiedlichen Hybridisierung mit den verschiedenen Materialen — auf der einen Seite das ferromagnetische Fe, auf der anderen das antiferromagnetische Mn — wurde für dieses System ein negativer Magnetowiderstand beobachtet.

Für die Untersuchungen der magnetischen Eigenschaften eines Molekülspins wurden Chromacetylacetonate $(Cr(acac)_3)$ verwendet (Kapitel 6). Diese bestehen aus einem zentralen Cr^{3+}-Ion, welches von drei Acetylacetonatgruppen $(C_5H_7O_2^+)$ umgeben ist und einen metallorganischen Komplex bildet. Die Moleküle besitzen einen entarteten Spin-$1/2$-Grundzustand [25]. Für die Moleküle auf einer Cu(111)-Oberfläche beobachtet man, dass sich deshalb eine Kondo-Resonanz ausbildet. Dies hat zur Folge, dass die $Cr(acac)_3$-Moleküle keinen stabilen Spin mehr besitzen. Dabei ist die Kondo-Resonanz nicht nur auf das Molekülzentrum beschränkt, sondern sie tritt auch in der Umgebung der Kondo-Störstellen auf. Ähnlich wie für Quantenspiegel [26] lässt sich dies anhand der Einflüsse des Oberflächenzustandes des Cu(111) erklären. $Cr(acac)_3$-Moleküle lassen sich also nicht für magnetische Bits verwenden, stattdessen bilden die beobachteten Ergebnisse eine Grundlage für eine weitere Untersuchung verschiedener Interferrenzeffekte der Oberflächenzustände.

1 Introduction

In the last decades, remarkable progress has been made in semiconductor industry towards the miniaturization of electronic devices following Moore's and Kryder's law [27, 28]. Since the modern information society needs increasing computing power and storage capacity accompanied by a cheap production, a lot of efforts are put into this development. Most of today's applications are based upon semiconductor technology. The recent developments were mainly achieved by optimizing geometries and downsizing existing devices. The physical limits of this top-down approach will be reached in the foreseeable future since devices of the size of single atoms or molecules will be required. Thus, the understanding of the electronic and magnetic properties on the nanometer scale plays an important role for future applications. Since devices on the molecular level introduce new physics, this field attracted significant interest after the theoretical prediction of devices constructed out of single molecules made by Aviram and Ratner in 1974 [7]. Although intense research was done in this field in the last years, there are still many problems which have to be solved on the way to future applications.

In most of modern hard disks, read heads based on magnetoresistance effects are already present. Starting with the discovery of the giant magnetoresistance (GMR) in 1988 [1, 2] and its application in hard disks, storage densities increased rapidly and enabled cheap hard disks with large storage capacities. This is a prominent example of the fast incorporation of fundamental research in industrial applications [29]. In the last years, devices based on the tunneling magnetoresistance (TMR) effect [30] outperformed the GMR devices [4, 5]. The TMR based components are facing another problem with downsizing: due to their high areal resistances, the currents for very small devices become too low for high frequency applications. However, in the GMR case, the magnetoresistance signal becomes too small. The latter problem can be addressed by decreasing the size of the devices and simultaneously keeping the current densities and the magnetic sensitivity on a sufficient high level [31]. Moreover the application of GMR based devices is not only restricted on read heads.

They are widely used in different sensors for example in automotive or biological applications [32–34].

A promising idea on the way of downsizing the magnetoresistance devices is to apply GMR building blocks based on single molecules. There are several studies focusing on the magnetoresistance of systems partly consisting of organic material [35–38] even on the single molecular level [39]. But all of these measurements were performed in the tunneling regime through organic films facing the same problems of high areal resistances. This issue shall be focused in this work, making the transition from the TMR to the GMR regime for single molecules.

The application of single molecules in data storage devices is not restricted on components based on magnetoresistance effects. Increasing of the storage density by downsizing the individual bits faces the problem of decreasing stability. Thus, the application of single molecular magnets is an auspicious approach [40, 41]. These molecules consist of magnetic ions carrying a spin serving as magnetic bit and a surrounding organic ligand. The organic part is destined to stabilize the spin of the ion and shield it against external influences. For application as magnetic bits, they have to be placed on a surface. Thus, they have to be studied when they are in contact with the surface, since the adsorption strongly influences their properties [42]. The scanning tunneling microscope (STM) [43] is a powerful tool to address both interesting aspects of molecular magnetism. It was introduced as one of the best methods to study the electron transport on the molecular level [9]. The STM tip is used to contact the molecule which is adsorbed on the surface. By doing this, a molecular junction is formed, which can be used to determine the conductance properties of individual molecules. For studying magnetic properties of surfaces or adsorbed nanostructures, the STM offers the possibility of using magnetic tips [15] as well as applying inelastic tunneling spectroscopy to study magnetic properties [21, 22].

This work combines the different STM techniques for the study of different properties of individual molecules. At first, in chapter 2, an introduction to the magnetic effects, which are important for the following work, and to STM is given. The experimental setup as well as the samples used in the experiment are presented in chapter 3. Phthalocyanine molecules, widely studied in different systems and configurations [24], were used to establish the molecular transport measurements. By changing the surfaces and the metal cations inside the molecules different influences on the transport were studied (chapter 4). With applying spin polarized STM in the transport measurements, it was possible to measure the magnetoresistance of an individual molecule in the contact regime shown in chapter 5. By this the first successful measurements of the GMR of an individual molecule could be measured. The groundbreaking success combining relatively large conductances with a high GMR ratio offers the huge possibility of contributing to downsizing of data storage devices. In chapter 6, a second class of molecules — the acetylacetonate — were used to study the magnetism of a metal ion shielded against the surface by organic ligands.

The theoretical DFT calculations supporting this experimental work have been performed by Ferdinand Evers and Alexej Bagrets of the Institute of Nanotechnology of the Karlsruhe Institute of Technology.

2 Experimental and theoretical background

The basis of research on molecular spintronics is to have the knowledge of fundamental properties. For spintronics, this is especially the magnetism of the different parts. Here the electrodes of the molecular junction play an important role. After understanding the magnetism of the bulk, magnetoresistance effect become pivotal when transport between ferromagnetic leads is examined. But for metal-organic molecules, the magnetism is not limited to the electrodes: the metallic part can cause different interesting effect as for example the Kondo effect. Thus, the present chapter will start with an introduction to magnetic effects, which are important for the present work. To address the different properties the scanning tunneling microscope (STM) with its various operational techniques has proven its high impact on modern research. The STM is the method of choice for studying the magneto-transport on individual molecules. The basic working principles and the different techniques will be described in the second part. The chapter completes with an introduction to molecular electronics and its impact on optimizing modern devices. The different methods to study molecular properties — especially with the STM — will be presented.

2.1 Magnetism

The phenomenon of magnetism is already known since 4000 b. c. The most famous application is still the compass for the orientation in the magnetic field of the earth. Since Gilbert started the scientific research in the field of magnetism in the 16th century, many additional applications of magnets have been introduced. In modern information technology, magnetism plays a remarkable role especially in the character of spintronics, combining electronics with spin effects. Two main discoveries, which are fundamental for this work, strongly affected the evolution of data storage devices: today's storage densities in hard disks — exceeding $350\,\mathrm{Gb/in^2}$ [3] — would be impossible without using one of two fundamental magnetotransport effects: the tunneling magnetoresistance (TMR) [30] and the giant magnetoresistance (GMR) [1, 2]. Both effects describe the transport between two magnetic leads separated by a spacing layer. The difference between the two effects is related to the nature of the spacing layer. While for the TMR the transport takes place via tunneling between the electrodes over a non-conductive layer, the GMR is dominated by interface effects in a conductive regime. A theoretical understanding of the ferromagnetism of the leads is essential and should be illustrated in the first section. Afterwards, an introduction to the TMR and GMR will be given.

2.1.1 Theoretical description of ferromagnets

The first theory focusing on ferromagnetic materials, namely the Weiss model, used localized moments to describe the spontaneous magnetic order in ferromagnetic materials [44, 45]. For rare earth metals this model is a good description because in their case the magnetism is dominated by the f electrons, localized around the atomic nuclei. For transition metals, magnetism originates from the d electrons, which are delocalized over the crystal and thus forming the conduction band. Motivated by the discrepancy between the Weiss model and the ferromagnetism of transition metals as well as the introduction of the band models for electronic states in metals [46], Stoner proposed in the 1930s his model based on three postulates [47, 48]:

- The d electrons are responsible for the magnetism.

- The electrons have to fulfill Fermi statistics.

- The exchange interaction between the spins is treated as a molecular field contribution in the Hamiltonian.

The basis of the Stoner model are two paramagnetic electron bands which are identical for both spin directions (see Fig. 2.1 (a)). In an external field, the bands are split for the two opposite spin channels due to the energy shift caused by the Zeeman energy. As the Fermi energy for both channels must remain on the same level, the occupation numbers depend strongly on the spin channel.

For ferromagnetic materials, the interactions between the electrons are described in terms of a molecular field contribution to the energy. The molecular field results in an band splitting of

$$\varepsilon_M = -k_B \theta_M \zeta,$$ [2.1]

where $\theta_M = \mu_B N M_0 / k_B$ and $\zeta = M/M_0$ are the characteristic parameters. When the occupation numbers

$$n_{\uparrow/\downarrow} = \frac{n}{2}(1 \pm \zeta)$$ [2.2]

are compared with the occupation numbers derived from the density of states (DOS) the equilibrium state is given in dependence of ζ:

$$\frac{k_B \theta_M}{\varepsilon_F} = \frac{1}{2\zeta}\left[(1+\zeta)^{\frac{2}{3}} - (1-\zeta)^{\frac{2}{3}}\right].$$ [2.3]

The result can be divided in three regimes:

1. $k_B\theta_M/\varepsilon_F < 0.67$: the system is non magnetic since ζ vanishes

2. $0.67 < k_B\theta_M/\varepsilon_F < 0.79$: the system is weak ferromagnetic

3. $0.79 < k_B\theta_M/\varepsilon_F$: the system is strong ferromagnetic since $\zeta \approx 1$.

With introducing the DOS at the Fermi edge, the condition for spontaneous magnetization can be rewritten to the well known Stoner criterion

$$\frac{2}{n}N(\varepsilon_F)k_B\theta_M \geq 1.$$ [2.4]

This means that a material becomes ferromagnetic when the either the DOS at ε_F or the molecular field term, which is a characteristic material property, becomes large.

If the Stoner criterion is fulfilled an internal exchange field causes the band splitting (see Fig. 2.1 (b)). Hence, filling the bands to the Fermi energy results in an spontaneous magnetization of the system, since in the so-called majority band there are more states to be filled. This finally results in a net magnetization of the itinerant electrons and can explain the finding that the transition metals show magnetic moments that are non-integer in units of the Bohr magneton.

Since in the Stoner model the conduction band of a ferromagnetic material carries spin, the charge transport is always combined with spin transport. The states around the Fermi edge of the system dominate in the electron transport. Therefore, the important quantity for spin transport is not the magnetization of the system but the spin polarization at the Fermi edge. Under certain conditions the transport can mainly be conducted by the minority channel. The DOS sketched in Figure 2.1 shows an exemplary case where the spin polarization is of opposite sign than the magnetization: there are more states at the Fermi energy for the minority channel.

2.1.2 The tunneling magnetoresistance

In general, magnetoresistance describes the transport across a device in dependence of the magnetic field or the magnetic configuration of the components. In the case of TMR and GMR it is the transport across a non-magnetic spacer sandwiched between two ferromagnetic leads. In both effects, the resistance depends on the relative magnetization of the electrodes but it relies on a different conduction mechanism. For the TMR it is an insulating spacing layer, that separates the two electrodes (see Fig. 2.2). This means that electrons are transmitted between the two electrodes via tunneling as described in more detail in section 2.2.1. During the tunneling process, the spin of the electrons is conserved. In case of parallel magnetized layers the electrons from the spin-up band of the source tunnel into the spin-up band of the drain and vice versa. For small voltages only the states at the Fermi edge contribute to the tunneling. A channel with a higher spin polarization

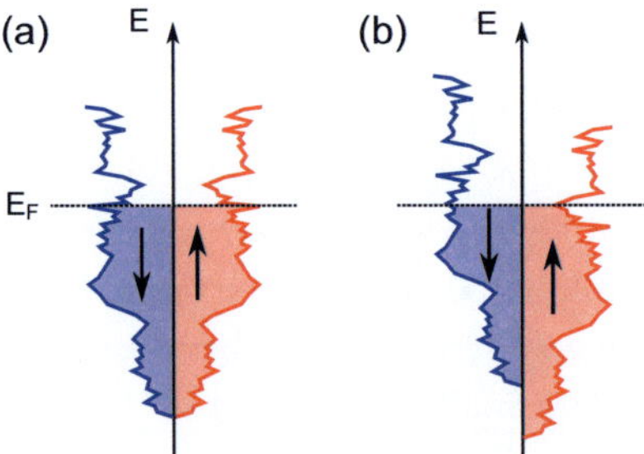

Figure 2.1: (a) Schematic density of states (DOS) picture for the two spin channels of a paramagnetic material without external field. (b) The bands are split for ferromagnetic materials. The number of majority ($\uparrow$) electrons is higher than that of the minority ($\downarrow$) ones

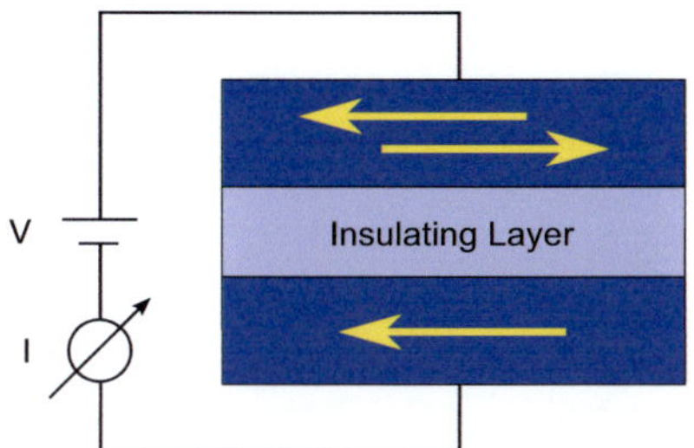

Figure 2.2: Schematic image of a magnetoresistance device. The magnetization of one of the magnetic leads can be reversed and the current is measured in dependence of the relative orientation.

at ε_F provides more electrons that can tunnel and more states to tunnel into. Since the number of electrons that are transmitted across the junction is proportional to both quantities, the transport strongly depends on the spin polarization. For a symmetric tunneling junction, the DOS for the analogous spin channels are the same in the parallel case. Thus, the channel with the high DOS offers many electrons at the source and many unoccupied states at the drain. This leads to a high contribution to the electron current, which is of course much higher than the contribution of the second channel with a low DOS at the Fermi edge. In the antiparallel case the spin channels are crossed and hence there is in both channels one electrode with a small DOS at ε_F. Therefore, the electron transport is quenched (see Fig. 2.3 (b)). Overall, the complete conductance for the parallel case is higher. This effect is the so-called tunneling magnetoresistance. The TMR ratio which is a measure for the strength of the effect can be defined in three different ways:

- optimistic TMR:

$$\text{TMR}_{\text{opt}} = \frac{G_> - G_<}{G_<}$$

- pessimistic TMR:

$$\text{TMR}_{\text{pes}} = \frac{G_> - G_<}{G_>}$$

- Asymmetry:

$$A = \frac{G_> - G_<}{G_> + G_<}$$

Here $G_>$ and $G_<$ stands for the conductance of the spin configuration, for which its value is higher or lower, respectively.

In 1975, Jullière was the first to report the TMR effect for ferromagnetic Fe and Co films separated by a non-magnetic Ge spacer layer [30]. He also connected the TMR ratio to the polarizations of the electrodes:

$$\Delta G / G = 2 P_1 P_2 \qquad [2.5]$$

with the spin polarizations

$$P_i = \frac{D_i^{\mathrm{maj}} - D_i^{\mathrm{min}}}{D_i^{\mathrm{maj}} + D_i^{\mathrm{min}}}.$$

and G being the average of the two channels. He reported an experimental TMR ratio of $\sim 14\,\%$ measured at $4.2\,\mathrm{K}$. In 1989, Slonczewski generalized Jullière's model for arbitrary angles between the magnetization of the electrodes [49]:

$$G(\theta) = G(1 + P_{\mathrm{tip}} P_{\mathrm{sample}} \cos \theta) \qquad [2.6]$$

For two antipodal magnetizations ($\theta_1 = 0$, $\theta_2 = \pi$) Jullière's formula is reproduced. In the 1990s, after the discovery of the GMR (see sect. 2.1.3) and its application in hard disks, there was an high effort in optimizing MR ratios not only for GMR devices but also for TMR devices. It begun with insulating spacing layers consisting

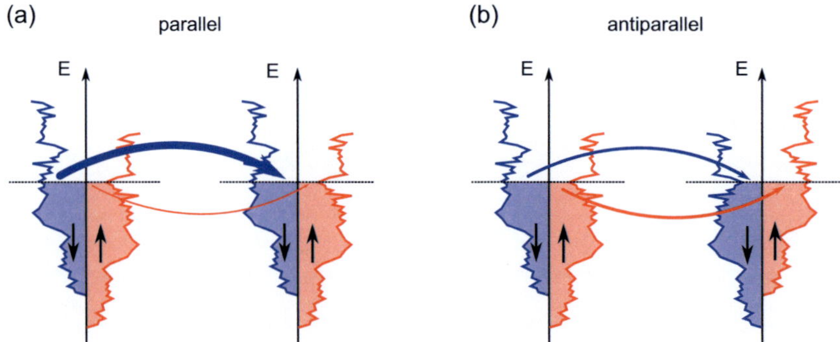

Figure 2.3: Schematic description of the TMR: (a) in the parallel case the minority channel has a strong contribution while the majority channel is quenched. (b) in the antiparallel case either the source or the drain states are reduced. Thus, the conductance is smaller than in the parallel case.

of Al_2O_3 revealing TMR ratios of $\sim 12\%$ at room temperature [50, 51]. For applicaton in read heads of hard disks, the Fe/MgO/Fe devices showed the highest TMR ratio ($\sim 200\%$ at room temperature) with a good technologic applicability [4, 5]. The crucial point is that the transmision across the MgO layer strongly depends on the symmetry of the electron states and thus the spin of the electrons. Due to this high MR ratios the MgO barriers are used in many modern hard disk read heads, outperforming GMR read heads. Nowadays, with further downsizing the magnetic domains of a hard disk, the TMR faces another problem. Due to the high areal resistance of the tunneling junction the currents become too low for sufficient small TMR junctions. Consequently, it is impossible to use such devices in high frequency applications due to shot noise and impedance mismatch [52].

2.1.3 The giant magnetoresistance

The GMR is a second magnetoresistance effect which is closely related to the TMR described in the previous section. The effect was simultaneously observed in the groups of Peter Grünberg and Albert Fert in 1988 [1, 2]. The first experiments have been performed on Fe/Cr/Fe systems representing two ferromagnetic layers separated by a antiferromagnetic spacing layer. Since the spacing layer in case of the GMR is conductive, the characteristics of its material have a stronger influence. However, for GMR junction the main influence are the interface properties. The hybridization between the different states of the two materials connected at the interfaces generally depends on the spin channel. For example the majority states of the leads hybridize much better with the spacing layer. Therefore, the transmission of the electrons across the interface is much higher for this channel. For the second channel exhibiting a lower hybridization the reflectivity is higher and thus the transmission is quenched.
For a parallel alignment of the electrodes (see Fig. 2.4 (a)), this means that the transmission over both interfaces is high for the majority spin channel. Hence, the electrons can be transmitted ballistically over the whole device and the resistance is small even though the minority channel does not contribute to the transport. In the antiparallel case (see Fig. 2.4 (b)), the states of the spacer layer hybridize with the majority states of each ferromagnetic electrode individually. Due to the antiparallel alignment the hybridization is not across the whole structure. This results in a

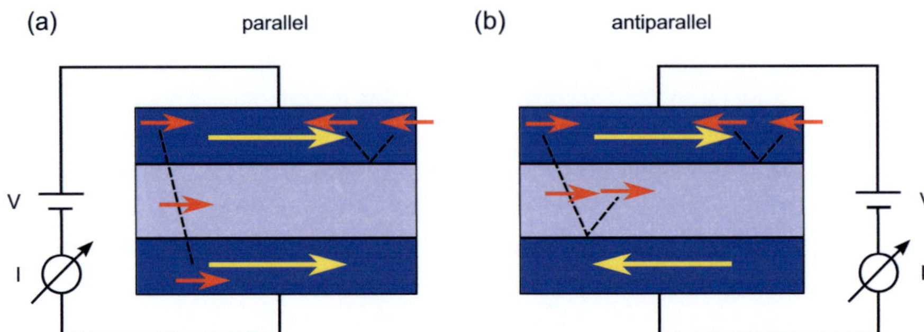

Figure 2.4: Sketch of the GMR for the two relative magnetizations. (a) For parallel alignment of the electrodes, the same channel hybridize strongly with both electrodes. The electrons are transmitted over both interfaces. (b) For the antiparallel alignment, the electrons are reflected either at the first or second interface due to weak hybridization.

strong reduction of the transmission for the antiparallel alignment [1,2].

The GMR had a high impact on the data storage technology allowing a large increase of the storage densities of hard disks. In 1997, IBM produced the first commercial hard disk using GMR. The high impact on the research field of spintronics resulted in awarding the Nobel prize in 2007 to Grünberg and Fert. For a long time the main research on the field of magnetoresistance devices was to increase the MR ratios. With the discovery of the high TMR ratios in MgO junctions [4, 5], their values exceeded the GMR ratios. Therefore, the TMR became more frequently implemented in hard disks. With today's domain sizes, GMR junctions become again more important. Due to their lower areal resistance the currents for TMR junctions become markedly small.

2.1.4 The Kondo effect

For individual spins that are adsorbed on a surface, different effects influences the magnetic properties. One of the important effects that have been observed for magnetic molecules is the Kondo effect [53]. Originally, it describes a low temperature anomaly of the resistance in metals with dilute magnetic impurities. It was discovered by de Haas and van den Berg in 1936 [54]. It took some time until it could be explained theoretically by Kondo in 1964. In his calculations he showed that an additional spin term in the Hamiltonian is responsible for this anomaly [55]. The increase of the resistance at low temperatures can be explained in the formation of the so-called Kondo resonance at the Fermi edge causing a higher probability of the conduction electrons to scatter from the impurity [56].

With the discovery of a closely related effect in transport measurements across quantum dots [57–59], the Kondo effect experienced an important comeback. Also for quantum dots, a Kondo resonance can be formed that is no longer responsible for a decrease of the conductance but provides a channel for the electron transport across the quantum dot. This results in a zero bias conductance [60, 61]. For the formation of the Kondo resonance, several conditions have to be fulfilled: a degenerated ground state, the possibility to switch between the ground states by transfer for example the spin to a lead electron, and strong coupling to the leads.

The Kondo resonance can be directly addressed by measuring the DOS of the System. The Kondo resonance appear as a Fano shaped peak at the Fermi edge. In order to determine the characteristic Energy scales the peak can be fitted by the Fano function:

$$\frac{\mathrm{d}I}{\mathrm{d}V} = A \cdot \frac{\left(\frac{eV+\varepsilon_0}{\Gamma} + q\right)^2}{1 + \left(\frac{eV+\varepsilon_0}{\Gamma}\right)^2} + B. \tag{2.7}$$

Here ε_0 is the peak position with respect to the Fermi edge, Γ is half width at half maximum of the peak, which is connected to the Kondo temperature

$$\Gamma = k_B T_K \tag{2.8}$$

and describe the caracteristic energy scale of the Kondo effect. A is the amplitude of the peak and B the background, are fitting parameters. q describes the type of the Fano resonance. For small q the Fano resonance appears as a Lorentzian dip, while for large q as a Lorentzian peak. q is directly connected to the coupling to the tip: a large q means a strong coupling to the Cr ion, while a small q represents a strong coupling to the surface.

Since its discovery in quantum dots, the Kondo effect was detected in various different nanostructures, for example in carbon nanotubes [62] or semiconducting nanowires [63]. In molecular transport the effect was discovered in individual divanadium molecules [64] and in C_{60} molecules [53]. The latter showed the first observed Kondo effect of a system with spin 1, which can be explained by an underscreened Kondo effect [65]. Also for phthalocyanine molecules, this effect can be observed after manipulating the organic ligand and thus the coupling to the substrate [66]. Another interesting application of the Kondo is the use in spectroscopy to determine spin states in molecular magnets [67]

2.2 Principles of the scanning tunneling microscope

Since its invention in 1982 [43], the use of scanning tunneling microscopy has strongly contributed to the study of numerous properties of various systems. Its boost on surface science resulted in awarding the Nobel prize to its inventors Binnig and Rohrer in 1986. As all other scanning probe techniques, the STM usually consists of a probe, *i.e.* the STM tip, built of a conductive material shaped as a sharp tip and a sample, which in the case of STM must be conductive. Between tip and sample a bias voltage is applied over the vacuum barrier separating them. This results in an effective current utilizing the tunneling effect which is the pivotal principle of the STM. After giving an introduction to the tunneling effect, the different working principles of the STM will be presented. In particular for studying magnetic properties this are the different spectroscopy techniques and the spin-polarized STM.

2.2.1 Theoretical description of the tunneling effect

The tunneling effect

The easiest way to explain the tunneling effect is to treat the electron as a quantum mechanical wave scattered at a finite potential barrier of the width a. The potential of this model system is determined by

$$V(x) = \begin{cases} 0 & x < -a, \\ V_0 & -a \leq x \leq a, \\ 0 & x > a. \end{cases} \qquad [2.9]$$

Classically, it would not be possible for the electron to be transmitted from one side of the barrier to the other as long as its energy is lower than the barrier height V_0. For a quantum mechanical treatment, one has to solve the Schroedinger equation for this system. The resulting wave function has the form:

$$\psi(x) = \begin{cases} e^{ikx} + re^{-ikx} & x < -a, \\ ae^{\kappa x} + be^{-\kappa x} & -a \leq x \leq a, \\ te^{ikx} & x > a, \end{cases} \qquad [2.10]$$

with $k = \sqrt{2m_e E}/\hbar$ and $\kappa = \sqrt{2m_e(V_0-E)}/\hbar$. At the barrier one part of the incoming wave e^{ikx} is reflected, represented by the part re^{-ikx}, while the other part te^{ikx} is transmitted to the other side of the barrier. Inside the barrier, the electron is described by an evanescent wave decaying exponentially over the barrier width. Using the four constraints for connecting the three parts of the wave function, the four constants a, b, r, and t can be calculated. This finally results in the transmission probability $T = |t^2|$:

$$\begin{aligned} T &= \frac{1}{1 + \frac{V_0^2}{4E(E-V_0)}\sinh(2\kappa a)} \\ &\approx \frac{4V_0^2}{E(E-V_0)}e^{-2\kappa a} \end{aligned} \qquad [2.11]$$

where the later simplification is valid for $e^{-\kappa a} \ll 1$, which is always fulfilled in the case of the STM.

Although this description of the tunneling effect is a quite simple approach it is useful to understand the basic principle of the STM. The transmission probability [2.11] and thus the tunneling current strongly depends on the tip-sample separation. Therefore, it can be used as a very sensitive measure for the distance between two conductive electrodes separated by a tunneling barrier which is in the STM geometry formed by the vacuum. This is used in the STM to map the surface of a sample by scanning it with a conductive tip and measuring the tunneling current. However, the remaining engineering challenge is to provide a stable system such that a very precise control of the movement of the tip is possible. Therefore, the system must be isolated against inner and outer vibrations that will otherwise highly influence the imaging quality. The control of the movement of the tip is done as for other scanning probe microscopes using piezo electric motors. Applying a voltage on these piezo electric elements leads to a controlled movement of either the tip or the sample. By using a two-stage system the coarse and the fine motion are separated to combine high range with high precision.

The naive operation mode of the STM is to use the so-called constant height mode. The tip is scanned at a constant z position above the sample while the tunneling current is recorded. Thereby, a map of the tunneling current in dependence of the x-y position of the tip is obtained. This allows to scan rapidly over the sample, while it always bears the risk to crash into the surface, when it shows a high deviation in the height compared to the tip-sample separation. Nowadays, it is more common to measure in the constant current mode. In this mode, the z position of the tip is recorded while the current is kept constant. This is achieved using a feedback loop, that reads out the current and adjusts the height of the tip accordingly. This slows down the possible scan speed, but helps to avoid tip crashes.

Bardeen's approach and Tersoff-Hamann model

Most of today's theories used to describe the tunneling process for STM are based on the approach presented by Bardeen in 1961 [68]. This theory was originally introduced to explain the tunneling in superconductor junctions. But its application is not constricted to this context it can be used for various other systems in par-

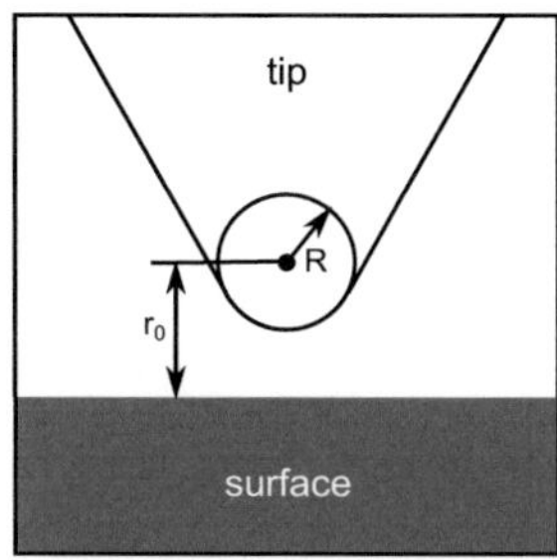

Figure 2.5: Schematic picture of the tunneling geometry in the STM used in the Tersoff-Hamann model. R characterizes the tip apex, r_0 is the tip-sample separation.

ticular for the STM geometry. The tunneling can be explained by a crossing from an occupied state μ of the source electrode to an unoccupied state ν of the drain, described by the tunneling matrix elements $M_{\mu\nu}$. For the correct description, one has to ensure mathematically that the initial states are occupied as well as the final ones are free. This is achieved by multiplication with factors of the Fermi function: $f(\varepsilon_\mu)$ or $1 - f(\varepsilon_\nu)$, respectively. The energy conservation can be ensured by a δ function. Thus, the tunneling current can be written as:

$$I_{\text{tun}} = \frac{2\pi e}{\hbar} \sum_{\mu,\nu} f(\varepsilon_\mu)[1 - f(\varepsilon_\nu)] \left| M_{\mu\nu} \right|^2 \delta(\varepsilon_\mu - \varepsilon_\nu). \qquad [2.12]$$

The remaining challenge is to calculate the elements of the tunneling matrix

$$M_{\mu\nu} = \frac{\hbar^2}{2m} \int d\vec{S} \left(\psi_\mu^* \vec{\nabla} \psi_\nu - \psi_\nu \vec{\nabla} \psi_\mu^* \right) \qquad [2.13]$$

where ψ_μ and ψ_ν are wave functions of tip and sample in the absence of tunneling. This means that the problem can be solved independently for tip and sample. However, further analysis can only be performed for known electrode geometries. The typical STM geometry is shown in Fig. 2.5. This model was at first used by Tersoff and Hamann to calculate the tunneling current in the regime of small voltages V and low temperatures T [69]. The foremost part of the tip has the strongest contribution to the transport. Tersoff and Hamann claimed that it can be described by a sphere with the radius R. Additionally the spherical s-type wave functions decay most slowly in vacuum. Thus, in first order only their contributions must be taken into account. With introducing the density of states of tip $\rho_{\text{tip}}(\varepsilon)$ and sample

$\rho_{\text{sam}}(\varepsilon)$ the tunneling current is given by

$$I_{\text{tun}} \propto \rho_{\text{tip}}(E_F)\rho_{\text{sam}}(r_0, E_F)V \exp\left(2R\sqrt{2m\phi}/\hbar\right),\qquad [2.14]$$

where $\rho_{\text{sam}}(r_0, E_F)$ is a handy writing for the local density of states (LDOS) of the sample at the position of the tip to combine the DOS of the sample with the transmission of the barrier:

$$\rho_{\text{sam}}(z,\varepsilon) = \rho_{\text{sam}}(\varepsilon)T(z,V,\varepsilon).\qquad [2.15]$$

For larger voltages, which are relevant for most experimental needs, [2.14] must be integrated over all the states that contribute to the tunneling current. This results in:

$$I_{\text{tun}} \propto \int_0^{eV} \mathrm{d}\varepsilon\, \rho_{\text{sam}}(E_F + \varepsilon)\rho_{\text{tip}}(E_F + \varepsilon - eV)T(z,V,\varepsilon).\qquad [2.16]$$

The transmission can be calculated in the WKB approximation [70]:

$$T(z,V,\varepsilon) \approx \exp\left\{-2z\sqrt{\frac{2m}{\hbar}\left(\bar{\phi} + \frac{eV}{2} - \varepsilon\right)}\right\}\qquad [2.17]$$

With this model one can explain most of the important features observed with STM. The proportionality of the tunneling current to the LDOS of the sample means that one measures a map of the LDOS rather than the real topography. For larger voltages, [2.17] results in an exponential dependence of the tunneling current on the applied bias. To get the actual LDOS this effect has to be corrected. It also shows that the DOS of the tip plays an important role on the images. For this purpose the usage of materials with a DOS as flat as possible around the Fermi edge, for example W, is recommended

2.2.2 Tunneling spectroscopy

But the STM is not restricted on measuring topographical images of surfaces. The dependence on the DOS also implies the idea of using the STM as a technique to measure it. This can be done by recording tunneling spectra. For higher energies inelastic properties become important, which also can be addressed with STM.

Elastic tunneling spectroscopy

To measure the LDOS of a sample the particular derivative dI/dV of the tunneling current $I(V)$ is taken at any point of the surface. For this purpose, the tip is kept at a constant height above the surface and the $I(V)$ curve is recorded by varying the bias. The best method to get dI/dV in a good quality is to use a lock-in amplifier. In this way the derivative is obtained by modulating the tunneling voltage with a periodic signal and integrating the resulting tunneling current. Thus, the resolution of the spectra is significantly higher than of the ones numerically derived from the bare $I(V)$ curves.

Since the Tersof Hamann approach and hence the direct proportionality of the tunneling current and the LDOS is only valid for small bias, the spectra measured for larger voltage ranges have to be corrected. The fastest method to eliminate the high voltage effects is to divide the differential conductance dI/dV by I/V. A more exact correction was introduced by Ukraintsev, which has to be used for spectra with even larger voltage range [70].

Inelastic tunneling spectroscopy (ITS)

So far, only the elastic tunneling process is contemplated in which the energy of the electron is conserved. When the energy of the tunneling electron exceeds certain thresholds, inelastic excitations can be created. This means that another channel opens for the transport, observable by a kink in the $I(V)$ curve both for positive and negative energy [71]. Since the contribution to the tunneling current is in most cases relatively small due to a low cross section in the creation of the excitations, it is hard to observe these kinks in the normal spectra. Therefore, it is recommended to measure either the first derivative dI/dV or even better the second derivative d^2I/dV^2 [72]. The opening of a new channel can be observed as steps in the dI/dV or

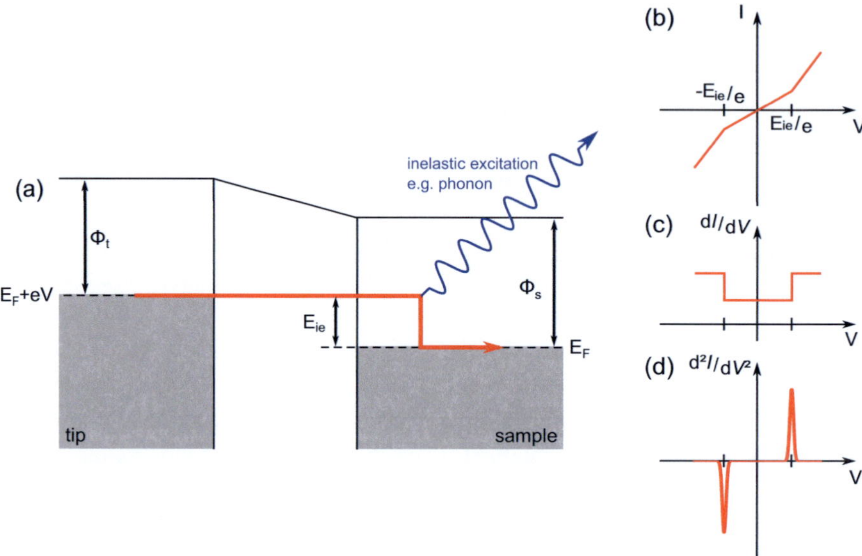

Figure 2.6: (a) Energy picture of the tunneling process. For low biases only elastic processes are possible. If the energy is high enough it can excite for example an phonon and relax to an unoccupied state. (b) This additional channel enhances the conductivity resulting in a larger slope of the $I(V)$ curve. (c) Since this effect is very small it is easier to be detected in the $\mathrm{d}I/\mathrm{d}V$ as a step or (d) as a pair of dip and peak in the $\mathrm{d}^2I/\mathrm{d}V^2$.

a pair of dip and peak in the $\mathrm{d}^2I/\mathrm{d}V^2$, respectively (see Fig. 4.10). ITS has been used to study various excitations like the phonon spectra of adsorbed molecules [21] or magnons in thin films [23].

2.2.3 Spin-polarized scanning tunneling microscopy

In section 2.1.2 the TMR was introduced as an effect that describes the tunneling in dependence on the relative magnetization between magnetic electrodes over a tunneling barrier. Since the normal STM geometry is a tunneling junction as well, one can directly think of using the TMR in a magnetic STM setup to study magnetic properties of surfaces [15]. In the STM setup, one of the electrodes is formed by the tip. To get magnetic images a magnetic tip has to be used. So far, three different methods are reported, which were successfully used to get magnetic tips for magnetic STM imaging:

- non-magnetic tips coated with magnetic materials like Fe, Co, Cr or Mn

- bulk magnetic tips, for example CrO_2 [15]

- coil tips which are built of a core consisting of a soft magnetic material with a low magnetostriction and a coil wound around the core. For these tips the magnetization is induced by a coil current [73].

The methods have different advantages and disadvantages. The coated tips can be prepared easily by deposition of magnetic material on ordinary wet-etched W tips. By using different materials or film thicknesses the magnetization of the tip can be varied from out-of-plane to in-plane orientation. Typically, it is difficult to control the film thickness accurately to get the favored properties in a controlled way. Another problem is forming the tip inside the STM without loosing the information about the actual magnetic properties of the tip.

These problems do not occur for bulk magnetic tips, while it is more difficult to prepare the desired tip shape out of the bulk wires. Also changing the properties of the tip from in-plane to out-of-plane magnetization costs much more effort, since an entirely new tip is needed.

For both kinds of tips, it is not necessary to use ferromagnetic materials as magnetic coating or bulk material. Since the tunneling is determined by the last atom it is also possible to use antiferromagnetic materials. The last atom which dominates the tunneling has still a well defined spin polarization. The latter have the advantage that the magnetic stray field of the tip is much smaller than for ferromagnetic

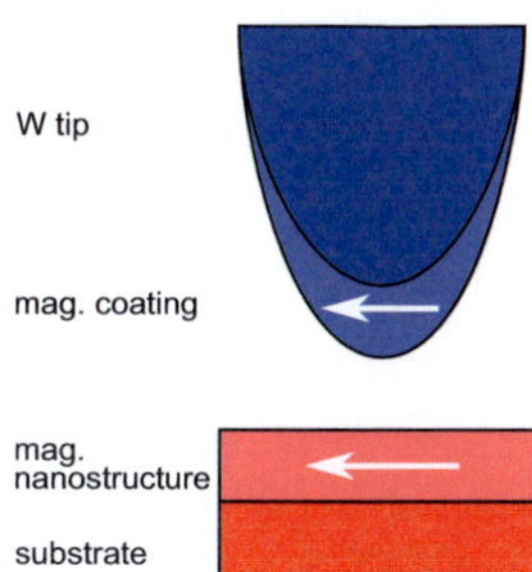

Figure 2.7: Sketch of a spin polarized STM. The tip consists of a W tip coated with a magnetic material. The surface and the tip should show the same anisotropy. Here, for example both are in-plane magnetized.

tips and the influence on the magnetic properties of the surface is weaker.

The magnetic properties can be controlled best for the coil tips, since the magnetization is completely determined by the induced field due to the current through the coil. A second big advantage of this method is the possibility to change the magnetization direction of the tip in an easy way by reversing the coil current. For this, tips with a very low magnetostriction are needed. However, it is very complicated to prepare the tiny coils necessary for applying the field. For out-of-plane sensitive tips, one has to fight very high stray fields. For in-plane sensitivity, ring tips are used [74]. In this case the stray field is not this problematic since for a closed ring the magnetic field lines are closed. For the connections of the coil, one needs a three terminal STM setup. This makes the tip transfer inside the UHV setup nearly impossible and reduces the measuring rate.

After implementing the tip as the first of the magnetic electrodes of the tunneling junction, information about the magnetic properties of the sample, serving as the second one, can be gained. The spin polarization of the surface strongly influences the tunneling current. Thus, it can be used to obtain information on the surface magnetism. The simplest method to extract the spin information is to use topographic measurements, as it was done in the first sp-STM measurements by Wiesendanger *et al.* [15]. Mono atomic step heights on the Cr(001) surface measured with a magnetic CrO_2 and a non-magnetic W tip are compared. While with non-magnetic tips all steps are of the same apparent height, the measurements with magnetic tips show an alternating step height. This can be explained by the layerwise antiferromagnetic structure of the Cr(001) surface. Since the spin dependent component of the tunneling current is quite small, this effect is hardly observable.

For a better separation of the topographic and the spin information, spin polarized tunneling spectroscopy is applied. Following the Stoner model, the DOS of the two spin channels are split resulting in a spin polarization at most energies. Thus, the derivative dI/dV at these energies depends strongly on the difference of the LDOS between the two spin channels [75]. By measuring spatial resolved dI/dV maps at different voltages, it is possible to gain information about the spin polarization of the different points of the surface at the according energies. Areas with the same polarization as the tip appear bright, the areas with the opposite one dark. For intermediate angles values in between are observed. This method is quite powerful and it was used to study the surface magnetism for various different systems.

The main problem of this spectroscopy mode is that there are still influences of non-magnetic effects. For example for nanostructures, growth geometries often play an important role which can also strongly influence the STS contrast (see for example [76]).

The coil tips can be used for a third measuring method. By periodically reversing the tip magnetization, the magnetic contribution to the tunneling current can be completely separated from other ones. Since the effective polarizations in [2.6] is reversed for opposite magnetization of the tip, the sum of opposite components results in the non-magnetic contribution while the spin information can be directly extracted from the difference [73].

2.3 Molecular electronics

The increasing need for cheap and fast data processing and storage devices of modern information technology requires smaller and smaller electronic building blocks. This ongoing optimization process will presumably reach the limit of downsizing existing components in the foreseeable future. Alternative concepts based on new physics become important in order to replace classical devices. One of these new promising approaches is to use organic materials to build parts of the devices. While for example for organic LEDs [77] or organic solar cells [78] the usage in commercial applications is already reality, it is still a dream for the future in other fields like in quantum computing or in data storage devices.

In the field of OLEDs the organic materials are mainly used to replace inorganic parts of the classical devices. This approach adds new properties, for example new electronic properties or the possibility to grow thin, flexible systems. Nevertheless, this approach does not introduce a lot of new physics on the way of replacing classical devices. But this is not the end of possibilities that are offered to the field of electronics by using organic materials. Modern chemistry allows to synthesize many different types of molecules. In combination with theoretical simulations, it is possible to design very complicated molecular systems that reveal properties designed for special needs. Since the chemical synthesis of the molecules is highly reproducible, all molecules have the same properties on a very small lenghthscale. Thus, building single devices out of such molecules could be a promising approach for replacing classical techniques. In 1974, Aviram and Rattner were the first proposing this idea. They made the theoretical prediction for the use of a single molecule to build a diode [7]. This proposition started a new field: the single molecular electronics.

2.3.1 Measuring the transport properties of single molecules

The main problem of studying different properties in the framework of single molecular electronics is to address these very small objects in the size of a few nanometers in a controlled manner. These problem exists for most cases of examined properties: from electronic transport to magnetism. When the theoretical proposition of single molecular devices came up in 1974, the experimental techniques were much behind. Hence, new methods to study these single molecular properties had to be developed at first. Finally, in the mid 1990s, experimentalists succeeded in establishing several techniques with distinct disadvantages and advantages. As the two main techniques prevailed the use of break junctions, where molecules are put in between a small nanogap, and the use of the STM, where molecules on a surface are contacted with the STM tip. Nevertheless, one should not forget about other methods used today. In the following, an introduction to the different methods should be given.

Break junctions

One possibility to measure the transport properties of single molecules is to produce a nanogap small enough to put a single molecule inside: the break junction method. A metallic wire, normally produced by lithography, is broken by pushing with a piezo actuator against the backbone of the substrate. By doing this, a small gap is created. Its width can be controlled by the piezo motors of the setup. The first transport measurements with the use of break junctions have been performed by Reed and co-workers. They used gold electrodes coated with thiol terminated molecules. The thiol groups have a high tendency to bind to the gold surface [8]. This should ensure the alignment of the molecules in the gap. The big disadvantage of this method is the lack of knowledge of what exactly happens inside the junction. To determine the molecular conductance a high statistic is needed in order to reduce the influence of the corrupt measurements with more than one molecule bridging the gap. By opening and closing the nanogap repetitively and measuring the conductance every time a high throughput is achieved. This leads to a widely spread histogram, that show steps at the conductance values of the molecule. In this way the molecular conductance is extracted.

A big advantage of the break junction method is the possibility to include a gate electrode on the backside of the substrate. This allows a three terminal measurement, more or less impossible with the STM. Via the gate voltage the molecular orbital can be shifted and hence forming a molecular single electron transistor [79].

Contacting with STM

In section 2.2, basic measuring techniques with the STM — topography plus elastic and inelastic tunneling spectroscopy — have been introduced. For studies of the molecular transport in the contact regime, these methods are not sufficient.Nevertheless, it is self-evident that putting a molecule into the tunneling gap seems to be a promising approach. Since the tip sample separation is controllable very precisely with the piezo voltage, the manifest idea is to approach the tip towards the surface atop of a molecule in a controlled way. If this experiment is performed on a bare substrate, an exponential increase of the tunneling current due to decreasing the tunneling barrier width is expected. The slope of the exponential curve is determined by the work function of the substrate. When the rip reaches the surface the exponential increase is stopped. In the point contact, the moment when the front most atom reaches the first vacuum layer of the molecule, one can observe a current of $\sim 1G_0$ where $G_0 = \frac{e^2}{\hbar}$ is the conductance quantum. This can be used to determine the initial tip-sample separation.

In 1995, Joachim *et al.* have been the first applying the method of taking current-distance curves atop of a molecule to determine its transport properties [9]. They studied the electronic transport through single C_{60} molecules. When approaching the tip atop of the spherical fullerene, the establishment of the contact with the molecule is observed when the exponential behavior fade to a constant one. The transition to the contact regime is preceded by an increase of the slope directly before forming the contact, indicating an attractive force between the tip and the molecule resulting in an elongation of the C_{60}. The current measured after reaching the constant regime now directly gives the molecular conductance.

Since then, more and more transport studies across single molecules have been performed with STM on various systems. C_{60} remains a widely used molecule for transport studies [16,17,19]. A big problem of the spherical shape of the fullerene is the smooth change from tunneling into contact regime, causing a high risk to crash into the surface. Haiss and co-workers used alkane chains with functionalized thiol groups as molecular wires and studied the transport with the STM [80]. These molecules show a jump into contact when the tip is approached, being a clear sign for establishing the molecular contact. This behavior simplifies the measurement procedure. Also the π conjugated perylenetetracarboxylicdianhydride (PTBDA) molecule shows this jumping.

The main advantages of the STM method is that by using topographic scans the alignment of the molecules at one of the electrodes can be studied very precisely before establishing the contact. Also the formation of the contact is much better controlled than in the case of using break junction. This reduces the statistic spread of the single measurements. Spacial resolved spectroscopies can image the molecular orbitals already attached to one of the electrodes, and hence study hybridization effects. Another advantage of the STM is the experience with sp-STM. This allows to perform spin dependent transport measurements. So far, this method has only been applied in the tunneling regime [39]. In this work the very first spin dependent transport measurements by contacting molecules with the sp-STM are shown.

Other methods

Most of the measuring methods beyond are closely related to these two methods. For using electromigration, which is a different method to produce a nanogap to deposit molecules on, high currents are applied in a controlled way on small nanocontacts normally produced by lithographic techniques [81]. By doing this the atoms are moved. This results in breaking the wire and forming the nanogap. Contacting of molecules with conducting magnetic atomic force microscopy tips [82,83] resemble the STM approach. A completely different approach is to assemble monolayers of molecules between two metallic, typically Au, electrodes and contact them via nanopores [13,84]. This approach has an even bigger problem ensuring the contact with a single molecule. This argument also holds for another method using a crossed wire geometry. A gold wire is coated with a self assembled monolayer (SAM) of molecules. Afterwards, it is put into contact with a second gold wire, measuring the molecular conductance of the SAM film [85].

2.3.2 Realized single molecular devices

Despite these big problems with measuring single molecular behavior, it was already possible to realize some electronic devices build out of single molecules. The simplest building block one can think of is just a conducting molecular wire, first realized out of benzene rings connected via sulfur bridges to gold electrodes [8,10]. The theoretical predicted molecular diode was realized for a SAM [86] and later for different single molecules [11, 12]. While the diode is the elementary building block for electronic circuits for data storage the application of molecular switches would be interesting. For various different molecules such switching behavior was reported [13, 14].

2.3.3 Magnetotransport in organic and molecular electronics

While magnetic transport measurements on single molecules are still very difficult, the vast knowledge of processing thin organic multilayers in combination with metallic interfaces allows to make magnetoresistance measurements on organic spacer layers [35]. A lot of effort was put into improving magnetotransport properties in organic junctions [36, 87–89]. Theoretical calculations for understanding the transport in molecular material were carried out down to the single molecular level [90, 91]. Some recent experiments have been performed on layers of copper phthalocyanine (CuPc) molecules [37]. Barraud *et al.* applied a new technique to measure magnetotransport in small ensembles of molecules [38]. But all of this measurements are facing two problems. On the one hand they are all carried out on ensembles of molecules not studying individual properties. On the other they are all performed in the tunneling regime. When thinking of an application in actual devices, these structures face the same problems in the downsizing process as the classical MR devices, for example too high areal resistance.

Some recent works study the magnetic properties of single phthalocyanine molecules on different surfaces using STM. Zhao and co-workers focused on the Kondo effect of CoPc on the non-magnetic Au(111) surface [66]. They were able to switch on the Kondo effect by manipulating the molecules with the STM tip. The combined experimental and theoretical work of Iacovita *et al.* studied the properties of the spin of the central atom of CoPc molecules on Co islands in the tunneling regime [39]. They imaged the spin of the central Co atom. To understand the hybridization of Pc molecules with the surfaces Atodiresei and co-workers have recently carried out calculations for molecules with delocalized π-systems and compared it to STM measurements on H_2Pc [92]. But so far, there are no reported measurements addressing the ballistic regime in the framework of magnetotransport on the level of single molecules. This will be done in this work.

3 Materials and methods

The study of such small nanostructures as individual molecules requires different conditions on the experimental setup. This is in particular the reduction of unwanted influences. The requirements have to be fulfilled by the experimental setup. To avoid errors due to perturbing adsorbates all the experiments are carried out in ultra high vacuum (UHV). Thermal fluctuations and drift of the nanostructures and the tip is reduced by carry out the experiments at helium temperature. External noise effects are reduced by damping the whole system. In the following the setup used for the present work will be introduced. This is not only the STM. Other techniques of surface science are used to ensure good reproducibility of the experiments. Subsequently, the different tips, substrates, and molecules that were used are presented. This includes the different preparation procedures. In section 2.3.1 the method to form molecular junction with the STM was introduced. The technique to extract the molecular conductance and thus the GMR ratio has to be improved to minimize the influence of the tunneling contributions. This method will be presented in the last part of the chapter.

3.1 Experimental setup

The opportunities provided by the STM to study objects of very small sizes have been shown in section 2.2 and 2.3.1. The high sensitivity of the STM presuppose different experimental conditions. In order to avoid crashes of the tip with the surface and to reduce noise effects an effective damping system is required. In the setup used in the present work this is achieved by a two-stage damping system. The whole machine is floating on four Newport I-2000 pneumatic insulators with laminar-flow damping, filtering out the external vibrations down to frequencies of $10\,\mathrm{Hz}$. Furthermore, the STM is mounted on four BeCu25 springs to damp internal noise down to $4\,\mathrm{Hz}$.

To avoid that the sample is contaminated with dirt and to ensure a controlled measurement, all the experiments were carried out in UHV: from preparation of the samples up to the STM measurements. Hence, the system has to provide different preparation facilities inside the UHV setup and has to enable the transport inside the setup. Thus, the setup consists of different UHV chambers that have to satisfy different needs. The different chambers are separated by valves and can be pumped individually.

The UHV inside the chambers is achieved by different pumping stages. A rotary pump is used to reduce the pressure to 10^{-3} mbar. With turbo molecular pumps (TMP) pressures down to 10^{-9} mbar can be achieved after baking the system to temperatures between $100\,°C$ and $250\,°C$ depending on the maximum baking temperature of the components. The base pressure of 10^{-11} mbar is achieved and kept by ion getter and titanium sublimation pumps. After baking, the TMPs are only used during sample preparation to reduce the contamination during sputtering or evaporation of the materials.

In Fig. 3.1 a top view of the whole setup is shown. In red, one can see the top part of the load-lock chamber that is used to insert samples into the UHV without breaking the vacuum in the two big chambers. This is done in two stages: the first chamber is pumped by a TMP down to 10^{-7} mbar, in the second chamber a getter pump reaches 10^{-9} mbar reducing the contamination down to a minimum. The two main chambers, namely the STM chamber (green) and the preparation chamber (blue), and their facilities are described in the following.

3.1.1 STM chamber

A third experimental condition, necessary for the present measurements, is that the STM is operated at low temperatures. This reduces the influence of thermal effects, especially the mobility of the molecules on the surface, the drift of the tip relative to the sample, and thermal fluctuations of the examined properties. Therefore, the STM is operated at a working temperature of around $4.2\,K$. This temperature is achieved by environ the STM by a liquid He bath cryostat. During measurement, the STM and the cryostat are decoupled and the STM is hanging on four springs to reduce the noise. During the sample transfer the STM is released from the springs and put into its parking position. Having thermal contact with the He

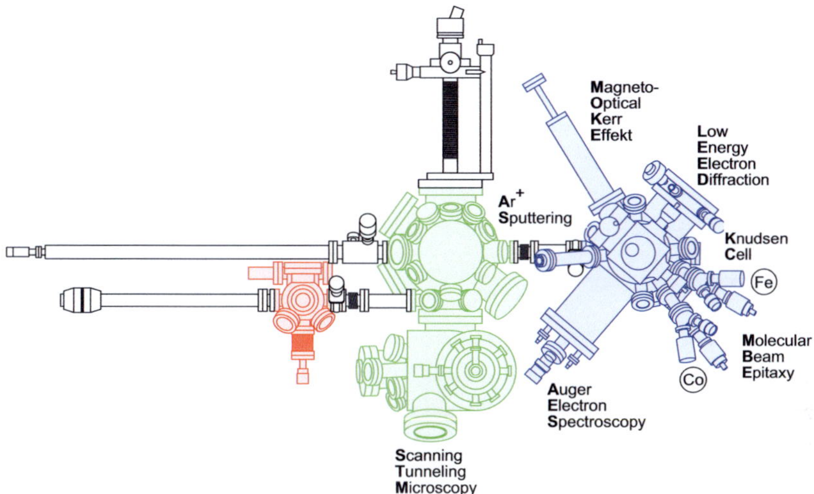

Figure 3.1: Top view of the STM setup used in this work. In red, the upper part of the load lock is shown; in blue, the preparation chamber with the different inbuilt components; in green, the STM chamber with cryostat. The black parts are the transfer rods and manipulators used to transport the samples inside the UHV chamber (modified after [93]).

cryostat quickens the cooling process for the sample. In order to reduce the Helium consumption and increase standing times, a second cryostat operating with liquid N_2 is mounted around the He one. While convection is avoided by having the cryostats in UHV, the input by thermal radiation is reduced by radiation shields coated with Al foil to enhance the reflectivity and thus reduce heat input. A door in both radiation shields allows to exchange the sample and tip without warming up the complete system, providing a higher operational capacity.

The actual STM is home-built, operated with commercial RHK electronics. The movement is two-staged: a coarse motion for all three direction is used to position the tip at the desired region close to the surface. The x-y motion is connected to the sample holder which is moved by different piezos in the two direction. The z motion is performed by a inchworm type piezo element connected to the tube scanner, which performs the fine motion during scanning. The tube scanner operates in all directions depending on where the voltage is applied. The coarse motion is actuated by the RHK PMC 100 while, for operating the measurements including the fine motion, the RHK SPM1000 is used. A computer with the RHK XPMpro software controls the electronics and the different measurement techniques.

Besides the STM, there is a carousel inside the chamber used to store up to six samples in UHV and a manipulator, providing the possibility to flash the tip without moving it to the preparation chamber.

3.1.2 Preparation chamber

The second main chamber of the system is the so-called preparation chamber. It provides several different facilities for the preparation of various sample and the analysis of their quality. For samples and tip cleaning an ion gun is used to sputter them by accelerating Ar^+ ions towards the sample. The beam is focused on and scanned over the surface. Thereby, the top most atoms are hit by these high-energy ions and sputtered away. This includes especially the unwanted adsorbates, which are dirt, films, and nanostructures deposited for previous experiments. Since the process is not completely homogeneous and crystal atoms are sputtered away as well, the surface is normally very rough after the sputtering. To get a flat surface and heal out the roughness, the crystals are heated by electron bombardment from a filament permanently mounted on the backside of the sample. Since some of the impurities of the crystal can diffuse to the surface during the heating process, one needs to apply several cycles of sputtering and annealing when inserting the new crystal into the UHV.

To check the cleanliness of the surface, the preparation chamber provides an Omicron SpectraLEED, consisting of electron gun with an analyzer screen to perform surface sensitive low energy electron diffraction (LEED) measurements [94]. LEED images are taken to investigate the crystallographic structure of the surface. This allows to gain information on the roughness of the sample and since some of the impurities yield reconstructions, they can be identified in the LEED pattern. As a second facility an Auger electron spectrometer (AES) [95, 96] is provided in the preparation chamber consisting of a Omicron EFK125 electron gun, a Omicron CMA150 analyzer, and a Omicron NGE50 control unit. By taking Auger spectra the chemical properties of the surface can be analyzed. By comparing the recorded spectra with known ones the chemical composition of the surface can be determined. This allows to detect unwanted adsorbates remaining on the surface.

For deposition of metallic nanostructures and films, the setup offers three evaporators for molecular beam epitaxy (MBE). The target is heated by electron bombardment at high voltages. The metal atoms are sublimed, partly ionized, and evaporated onto the sample. Different metals can be loaded either as a metal rod or in a molybdenum crucible, depending on the specific properties of the material. To quantify the evaporation process the Auger gun and the LEED screen can be combined to measure medium electron energy diffraction (MEED) [97] oscillations: at every maximum of reflectivity a full layer is completed and by counting the maxima the deposition rates can be calibrated (see Fig. 3.2).

The molecular evaporator, a so-called Knudsen cell, works similarly: a Cu body is heated resistively. This leads to a relatively homogeneous sublimation of the molecules. Since the molecules are not ionized and thus not focused in a controlled way towards the sample, the thermal beam is spread over the sample. Hence, a big fraction is adsorbed not only on the sample but at certain parts of the chamber, like the gate valve or the walls of the chamber. Therefore, the surroundings of the Knudsen cell have to be cleaned regularly, especially when changing the molecules.

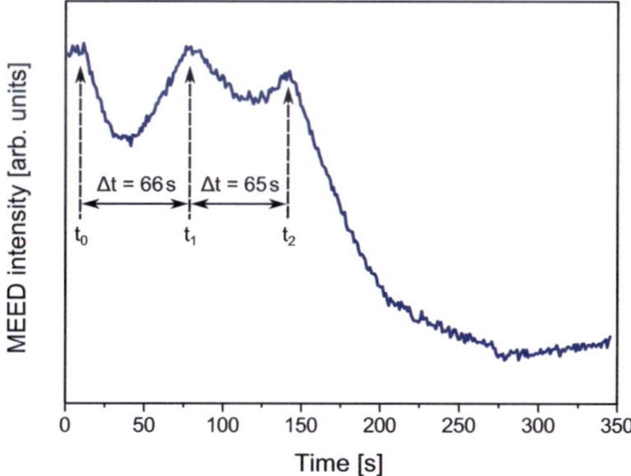

Figure 3.2: MEED graph taken for the growth of Mn on Fe(100) taken at a sample temperature of approx. 50 °C. The deposition is started at $t_0 = 9$ s. The first monolayer is completed at $t_1 = 75$ s and the second one at $t_2 = 140$ s. Afterwards, Mn grows three dimensional for this sample temperatures. One can extract a rate of 66 s/ML used for subsequent experiments.

The setup offers also the possibility to measure the magnetooptical Kerr effect (MOKE) of the samples [98]. It was used to ensure the magnetization and to measure the reversal fields of the Fe tip and the Fe whisker for the experiment shown in section 5.2.

3.2 Substrates and Tips

To perform magnetotransport measurements with the STM, magnetic electrodes are required. Since a sp-STM is used, this is a magnetic tip as well as a magnetic substrate on which the molecules are put. Further, for GMR measurements it has to be possible to measure in different magnetic configurations. Therefore, either the direction of the tip or the substrate magnetization has to be changeable. Since the coercive field of the tip is usually higher than the one of the surface, the latter can be reversed more easily. There are two different possibilities, how the reversal can be achieved. Either an external field is applied to change the magnetization of the sample or a substrate is used which provides intrinsically domains with opposite magnetizations. Reversing the magnetization by an external field has the advantage that it is possible to measure on exactly the same molecule with the same configuration for the two magnetization directions. Unfortunately, due to magnetostriction effects it is more or less impossible to change the external field while being in contact with the molecule. The most important advantage of working with structures forming surface domains is that you directly see your magnetic contrast by taking dI/dV maps. Since the spin contrast of coated tips is not *per se* ensured after preparation, this direct proof is very helpful. For this reason, two well studied substrates with opposite magnetic domains have been chosen: on the one hand Co islands on Cu(111), on the other Mn film on Fe(100).

Substrates

When a sub-monolayer amount ($\sim 0.5\,\mathrm{ML}$) of Co is deposited on a Cu(111) surface triangular shaped double layer islands are formed (see Fig. 3.3 (a)) [99]. A closer look into these Co island reveals that there are two different stacking types: one of the stacking of fcc type while the other is based on a stacking fault [100]. Due to this difference the triangle of the islands is either pointing upwards or downwards.

Besides the different topographies the stacking influences the electronic properties. Distinct states on the different islands were found [101]. Much more important for this work than the stacking difference is a second effect: Pietzsch and co-workers applied sp-STS measurements to study the spin polarization of these islands [76]. They found a spin polarization of the islands which is present over a wide energy range and dominates in certain energy regions over the stacking effect. The spin polarization is highest at around $-400\,$meV. This energy can be used to determine the magnetization of the islands and to find islands which are magnetized in different directions (see Fig. 3.3b). They also showed that, without applying an external field during Co deposition, the islands are magnetized out-of-plane. For both stackings the magnetization is arbitrarily either pointing out of or in to the surface. Due to this presence of islands with opposite spin polarization, this is an ideal system for GMR measurements. An additional advantage of this system is that there are areas with different surface materials.The deposited molecules can be adsorbed on the Co islands as well as on the bare Cu(111) substrate. This allows to study the influence of the electrode material on the transport within the one preparation.

As substrate a commercial Cu(111) crystal was used. After inserting the crystal into

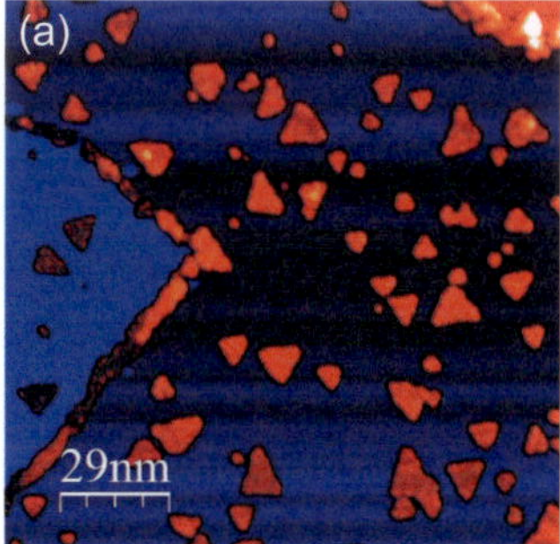

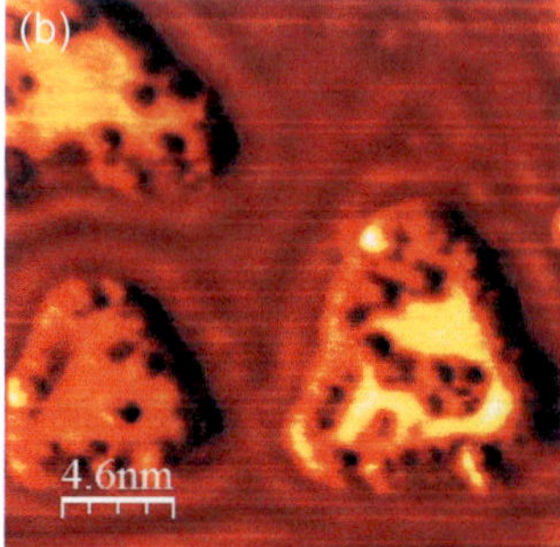

Figure 3.3: (a) Large scale scan of a submonolayer amount of Co deposited on Cu(111). The Co is forming double-layer islands. Depending on the stacking they are either pointing up or down. (b) Spatial resolved dI/dV maps taken with a magnetic tip reveal different magnetization of the islands. Bright yellow means parallel alignment of the magnetization of the island with respect to the tip magnetization, while a dark color indicates an antiparallel configuration.

the UHV system, the Cu(111) surface was cleaned by several cycles of Ar^+ sputtering and annealing. The sample was usually sputtered for $1-2\,h$ with an ion current of $\sim 3\,\mu A$. This ensures that the topmost layers were removed from the surface in order to eliminate unwanted adsorbates. For healing out the resulting roughness, the sample was heated by electron bombardment from the backside ($\sim 0.3\,°C/s$) until it reached a temperature of $450\,°C$. The sample was left for a short time at this temperature and subsequently cooled down slowly ($\sim -0.3\,°C/s$) to room temperature. This was repeated until the cleanliness of the sample was ensured in LEED patterns and STM images. On the clean sample a sub-monolayer amount of Co was deposited at room temperature. This leads to the formation of the desired Co islands.

A second system, which combines the advantages of reversing the magnetization and different surface domains, are Mn films on Fe(100). First measurements with spin polarized electron energy loss spectroscopy revealed an antiferromagnetic coupling between Mn layers grown on Fe(100) [102]. The coupling between the topmost Fe layer and the first Mn layer was disputed and could be connected to the oxygen contamination in the UHV equipment: while for a clean substrate it is ferromagnetic, it is antiferromagnetic when contaminated with O_2 [103]. First, sp-STM studies showed the different magnetization direction depending on the film thickness [104]. While the spin polarization is present over most parts of the energy range, the highest value is reached around $800\,meV$. It is important what happens when the Mn grows atop of step edges. Step edges both in the Fe and in the Mn result in a reversal of the spin polarization (see Fig. 3.4). While the latter can be easily seen in topographic images the first ones are hard to observe due to small differences in the atomic sizes of Mn and Fe (see Fig. 5.6). Schlickum *et al.* observed a domain wall of the Mn layers at these points, where the Mn atoms at the edge show frustration and form a domain wall depending on the film thickness [105]. The great advantage of this system is that besides the direct control of the magnetization at step edges, the Fe whisker and thus the spin polarization can be switched in an external field. The disadvantage of the Mn and Fe surfaces is that they are less noble than the Co and Cu surfaces and thus more dirt adsorbs during deposition of the molecules. This handicaps the detection of individual molecules for the transport measurements.

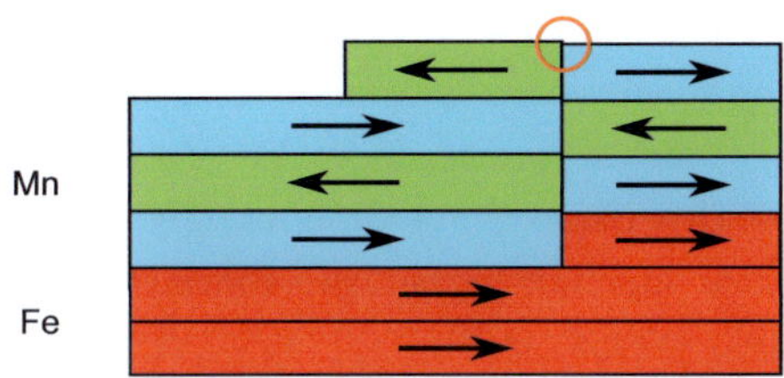

Figure 3.4: Schematic picture of the magnetization of Mn layers grown on Fe(100). For a clean surface the first Mn layer couples ferromagnetically to the Fe. The following layers couple antiferromagnetically. This results in different surface magnetizations depending on the number of Mn layers. An buried step edge can be seen in topographic images only by a small height difference stressed by the orange circle.

For the experiments, a home grown Fe whisker was used as substrate. To get a clean Fe surface a more intense preparation than for the Cu crystal is needed. Hence, longer sputtering cycles were performed while the sample was heated oscillatorily to temperatures between room temperature and 500 °C. This helped to get rid of oxygen contamination inside the whisker, since the oxygen atoms defuse to the surface at higher temperatures. A final healing was performed by heating to 450 °C. Since the whole procedure did not remove all contaminants, 15 ML of clean Fe was deposited on top and annealed only to 300 °C. This leads to a clean and flat Fe(100) surface. Previous experiments showed that the best sample temperatures for the deposition of Mn on Fe are between 150 °C and 200 °C [106]. At these temperatures 6–8 ML of Mn were deposited. Directly afterwards, the H_2Pc molecules were deposited to reduce the dirt adsorption on the surface.

Tips

For all measurements, wet-etched W tips were used. They were produced from highly clean W wires ($\varnothing = 400\,\mu m$) by etching with 5 % NaOH. This procedure ensures a very well shaped tip with a sharp apex and a stable enough body. After putting them in UHV they were cleaned by Ar^+ sputtering. Afterwards they were flashed with electron bombardment to high temperatures (800 V, 15 - 50 mA) for additional cleaning and forming of the apex.

For the spin polarized measurements magnetic materials were deposited onto the tip. Depending on the magnetization of the substrate, the material and parameters had to be adjusted to achieve the desired tip magnetization. For the Co islands being magnetized out-of-plane, an out-of-plane sensitive tip is needed. This was achieved by depositing 15 ML of Co onto the tip [39]. The use of Co also ensures a symmetric measurement with the same electrode material on both sides. For the in-plane magnetization of Mn, an in-plane sensitive tip is needed. Hence, 15 ML of Fe were deposited on the W tip, already used in former studies on Mn films on Fe(100) [104]. The Fe was slightly annealed after deposition (without applying high voltage, 5 mA) to produce bigger Fe domains and thus higher spin contrast.

3.3 Molecules

In section 2.3 we introduced various studies on different molecules. For the application in the present work, the molecules have to fulfill certain requirements. In order to look at magnetic properties the possibility to put different magnetic metals inside a metal-organic complex offers great opportunities. Furthermore, every exposure of the clean sample to air will lead to high contamination of the surface. Thus, it is essential to be able to deposit the molecules *in situ* under UHV conditions. Ideally, the molecules are evaporated by heating them with the Knudsen cell to their sublimation temperature without destroying their internal structure. Another important point for contact measurements is the molecular geometry. In section 2.3.1 we showed to different types of transition to the contact regime. For sp-STM measurements tip crashes must be avoided strictly. Thus the molecule's geometry should favor a sudden formation of a contact when the tip is approached. For the studies of magnetic properties of a shielded spin on a surface the molecule should ensure that the metal ion is separated from the surface.

Following these requirements two classes of molecules have been chosen: the phthalocyanines for the transport measurements and the acetylacetonates for studying magnetic properties of molecules on surfaces. They will be introduced in the following.

3.3.1 Phthalocyanine molecules

The phthalocyanine (Pc) molecules perfectly fit into the requirement profile. The molecules consist out of four aromatic isoindole subunits linked together by four nitrogen bonds and are thus forming a closed aromatic ring system (see fig. 3.5). This geometry ensures a high thermal stability and hence the possibility to deposit them in UHV. The Pc molecules are long known to be evaporable in UHV [107]. The two hydrogen atoms in the central cavity can be replaced by various different metal cations — in particular CoPc and MnPc used in this work.

The first and still most important application of the Pc molecules is as dye. Since its first description in 1907 [108] its brilliant blue shine was obvious and, thirty years later, it led to the commercial production of CuPc as artificial pigment. In 2000, 20 % of the dyes were phthalocyanines and their derivates [109]. However, their applications did not remain fixed on dyes. Due to their useful properties they became interesting for various purposes. Pc molecules are used in gas sensors [110] and as electron injection layers in organic solar cells [111]. Because of their usage in solar cells, especially the thin film behaviour is very well studied [112, 113].

The unique shape of the Pc molecules make them ideally suited for STM studies and CuPc had been one of the first molecules observed individually adsorbed on a surface with the STM [114]. Two years later, Lippel *et al.* succeeded in imaging the internal structure of CuPc [115]. The tunneling rates and charging effects of a quantum dot consisting of a single CuPc molecule was studied using STM measurements in the tunneling regime [116]. Magnetic properties have to be considered by putting magnetic atoms like Fe, Co, or Mn inside the molecule. While

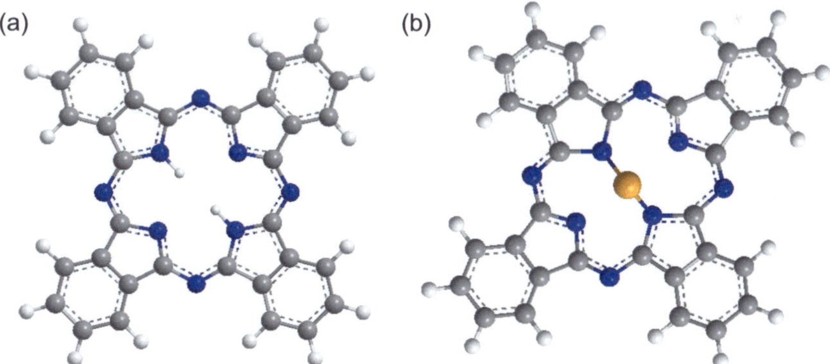

Figure 3.5: Model of the phthalocyanine molecules used in the experiments. (a) Hydrogen phthalocyanine, (b) metal phthalocyanine (C = grey, N = blue, H = white, Me = orange). While the H_2Pc is completely flat the MePc are slightly distorted due to the compound forming with the metal cation.

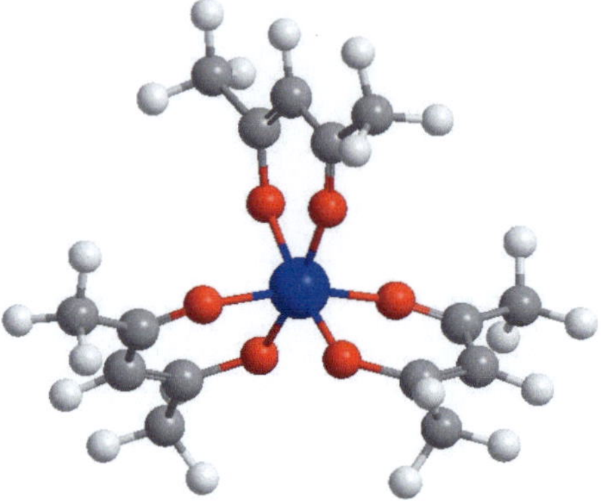

Figure 3.6: Model of the Chromium acetylacetonate. The three organic side-groups are aligned around the central Cr ion (C = grey, O = red, H = white, Me = blue).

CoPc molecules on the Au(111) surface showed a Kondo effect [66], it was possible to image the spin of CoPc on Co islands [39]. Calculations showed that the magnetism is highly dependent on the central metal atom [117].

In this work, the Pc molecules were deposited from the crucible in the Knudsen cell by heating them to 275 °C. A deposition time of around 1.5 min resulted in the requested coverage of the molecules on the surface. During deposition on the Co islands on Cu(111), the sample was cooled to 0 °C to reduce the surface diffusion of the molecules. Otherwise nearly all of the molecules were sticking to step edges or rims of the Co islands. For the case of MnPc, the deposition was done without cooling in order to reduce the adsorption of dirt on the highly reactive surface.

The molecules used in this work are commercially produced. Afterwards they are cleaned by repeatedly sublimating and condensing them by the group of Martin Bowen and Eric Beaurepaire at the IPCMS in Strasbourg.

3.3.2 Acetylacetonates

Since for the measurements of the magnetic properties of a molecular spin, the organic part of the molecules should provide an extended cage to isolate the cation from the surface, the Pc molecules are not suited for this purpose. Instead, the metal acetylacetonates (acac) have been chosen. These molecules consist of a central metal cation surrounded by a different amount of pentane-2,4-dione molecules depending on the oxidation number. In contrast to the phthalocyanines, the organic part does not form a closed organic system. Thus, the molecules are less stable and more difficult to evaporate without being degenerated. Due to this behavior, the acetylacetonates are widely used as precursor material in chemical vapor deposition of metal films. It was shown that $Fe(acac)_3$, $Ni(acac)_2$, and $Mn(acac)_3$ degenerate before reaching the evaporation temperature [118] and hence could not be used for *in situ* preparation in the UHV. In the same work it is reported that the $Cr(acac)_3$ and $Al(acac)_3$ compounds are stable and thus it should be possible to deposit the molecules *in situ* in the UHV [119]. Grillo and co-workers studied the dissociation of a $Cr(acac)_3$ derivative on a $Cu(100)$ surface with STM [120], but so far there have not been any studies focusing on the deposition and adsorption of these molecules onto the surface.

Since the $Cr(acac)_3$ molecules are evaporable and have a magnetic center, they were chosen for the experiments. The deposition of the molecules is more complicated than for the phthalocyanines. Since the molecules are less stable, due to the lack of a closed organic structure, the heating to the sublimation temperature of $90\,°C$ had to be done with low currents (max. $1.9\,A$) and therefore very slowly. After reaching the evaporation temperature the molecules are highly volatile and a deposition time of $1\,s$ was enough to get a quite high coverage of molecules.

3.4 STM-measuring method: obtaining current-distance curves

In this work the method described in section 2.3.1 was used to measure the electron transport across the phthalocyanine molecules. Topographical images were taken to position the tip at a certain position atop of the sidegroups of the single molecule. For the transport measurements, only clean and isolated single molecules were used. Furthermore, the molecules had to show the typical shape of the Pc molecules (see Fig. 4.1) to avoid degenerated molecules. After spatially positioning, the tip was moved towards the surface in controlled way ($\approx 1\,Å/s$), by applying a slowly increasing voltage to the z-piezo. The process is operated by the STM electronics. A voltage divider was used to reduce the approach speed and thereby measure curves with better resolution and reducing the risk to crash into the surface. During the complete approach procedure, the tunneling current was recorded. The maximum voltage, equivalent to the shortest distance between tip and surface, is increased slowly until a discontinuity in the current-distance curve could be observed. The discontinuity can be explained by the molecule switching to the contact position. this means the molecule is connected with one end to the surface and with the other one to the tip. The raw curves were modified by calculating the displacement of the tip from the set point position from the applied voltage using the piezo constant (0.93 nm/V) and the conductance in G_0 from the tunneling current

$$\frac{G}{G_0} = \frac{I}{V_{tun}G_0}.$$
[3.1]

Here, V_{tun} is the applied tunneling bias, which is constant during the complete approach procedure. For the Pc molecules the distance curves were recorded at biases of around 10 mV, since higher voltages lead to degeneration of the molecules due to high power input.

Since the tip apex is generally very broad compared with the relevant contact area, there are still contribution from the tunneling when the molecule forms the contact with the STM. Since there is no possibility to separate the two contributions from another, one has to use a model to reduce the error produced by the tunneling contribution. This is especially important for the GMR measurements where the TMR contribution should be eliminated completely. In Figure 3.7 the situation before

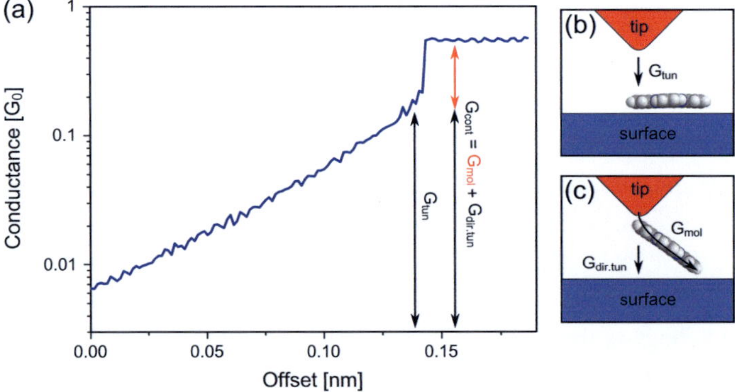

Figure 3.7: (a) Typical distance curve measured on Pc. Schematic images of (b) the tunneling geometry and (c) the contact geometry shall illustrate the way the molecular conductance is extracted: To reduce the influence of the tunneling contribution the conductance before forming the contact is subtracted from the one after.

and after establishing the contact is sketched. To extract the molecular conductance G_{mol} out of the measured conductance G_{mes}, it was assumed that the tunneling conductance G_{tun} before is of the same order than the tunneling contribution $G_{dir.tun}$ after forming the contact. This allowed to calculate the molecular conductance

$$G_{mol} = G_{mes} - G_{dir.tun} \approx G_{mes} - G_{tun}. \qquad [3.2]$$

This model is the same as already used in the work of Haiss *et al.*, in 2006 [18].

4 Non-magnetic transport properties of single phthalocyanines

For magnetotransport measurements on single molecules two very sensitive methods, *i. e.* contacting measurements with spin-polarized STM, have to be combined. Thus, preliminary work with non-magnetic tips was carried out. This helped to establish the technique at the present setup as well as to gain fundamental knowledge on the transport across the phthalocyanine molecules. These studies serve as basis for the GMR measurements.

In section 3.3.1, the phthalocyanine molecules were introduced as a good choice for transport measurements on single molecules with the STM. The growth and preparation of Pc molecules especially in thin film was widely studied due to their applications in organic solar cells [109]. The Pc molecules show a very low vapor pressure at room temperature allowing for an *in-situ* preparation under UHV conditions. This ensures the growth and further examination in a clean environment with a strongly reduced influence of impurities. By putting different metal cations in the center, their influence on the transport can be studied. Additionally, the geometrical shape of the Pc molecules offer a nice recognition value. Further their flat geometry should cause a sharp transition to the contact regime. To perform the transport measurements the molecules have to be adsorbed individually on the surface. The deposition was optimized to reach this goal by studying growth properties. On the individually adsorbed molecules of the optimized samples, transport studies were performed. By varying the molecular type (H_2Pc, CoPc, MnPc) and the surfaces (Cu, Co) distinct influences were examined. In order to explain the different observations — how the conductances depend on the metal cations and the substrates — tunneling spectra were recorded. Spatial resolved dI/dV maps were taken to visualize the molecular orbitals. Finally, inelastic tunneling spectroscopy was used to determine vibron excitations playing an important role for the process of contact formation.

4.1 Growth of the phthalocyanines

In order to use the samples for transport measurements on single molecules, it has to be ensured that the molecules are adsorbed individually on the surface. The molecules were deposited from a Knudsen cell as described in chapter 3.3.1. By heating them above their sublimation temperature, the molecules were sublimated and evaporated on surface.

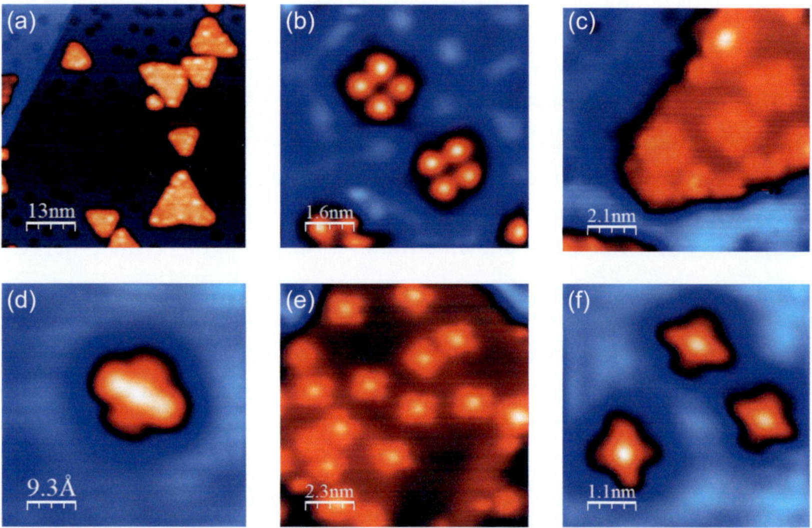

Figure 4.1: Topographical images of the different molecules on the surface taken at $T = 4.3\,\mathrm{K}$. (a) Overview scan of H_2Pc molecules deposited on Co/Cu(111) (100 mV, 100 nA). The single molecules lie flat on both surfaces. (b) Zoom onto two individual H_2Pc molecules on the bare Cu ($-10\,\mathrm{mV}$, 100 pA). The absence of a metal ion can be seen by the depression in the central cavity. (c) H_2Pc molecules on an Co islands (100 mV, 100 nA). At the edges of the Co islands, attached molecules are visible. (d) Individual CoPc on Cu (100 mV, 5 nA). The CoPc having only a two-fold symmetry is observed indicated by the brighter appearance of two of the side groups. (e) CoPc on Co (200 mV, 13 nA). The bright Co in the center can be used to distinguish between H_2Pc and CoPc. (f) MnPc on Cu (15 mV, 5 nA). MnPc also shows a bright center and a two-fold symmetry.

The measurements were performed for three types of phthalocyanines: H_2Pc, CoPc, and MnPc. This allowed to study the influence of the central atom on the transport across the molecule. The deposition on the Co islands on Cu(111) made it possible to examine the influence of the electrode material. Since the preparation of clean MnPc was rather difficult, these experiments were carried out only on bare Cu(111). In this way five different experiments could be performed: H_2Pc on Cu and Co, CoPc on Cu and Co and MnPc on Cu.

After deposition of the molecules, topographic scans were used to characterize the growth of the molecules. The molecules could be easily identified on the surface by their four aromatic isoindole side groups (see Fig. 4.1). Evaporation of the molecules with the sample at room temperature gave rise to surface diffusion resulting in the attachment of the molecules to the step edges and the rims of the Co islands. By cooling down the sample to $0\,^\circ C$ during deposition, single molecules were adsorbed on the Co islands as well as on the bare Cu. Figure 4.1 shows different STM topographies of the three types of molecules on the different surfaces. For all molecules a flat adsorption on the surface was observed. The metal free molecule can be identified by the absence of a bright area in the center, while for MnPc and CoPc the metal ion is visible in the center. The analysis of the adsorption angles of the H_2Pc molecules on Co gives clear peaks at around -85° and -55° plus some additional scatter around -25° (see Fig. 4.2). This distribution is consistent with a fourfold molecule adsorbed on the sixfold hexagonal Co(111) surface in a single adsorption geometry. To determine the exact adsorption site, a theoretical geometry optimization for H_2Pc on the Co(111) surface is modeled with a 65 atom cluster. The analysis suggests that H_2Pc adsorbs preferentially in the bridge position onto Co(111) owing to a binding energy $-8.17\,eV$ that is more negative than the ones found in either the hollow site position ($-8.06\,eV$) or the atop site position ($-7.45\,eV$). This is consistent with earlier findings [39, 121].

4.2 Influence of the central atom on the transport

The topographic images serve as a basis to perform transport measurements. The molecular transport is dominated by two main influences. On the one hand, the hybridization of the molecule with the electrodes determine the injection rates for the electrons into the molecule. On the other, the molecular orbitals of the molecule

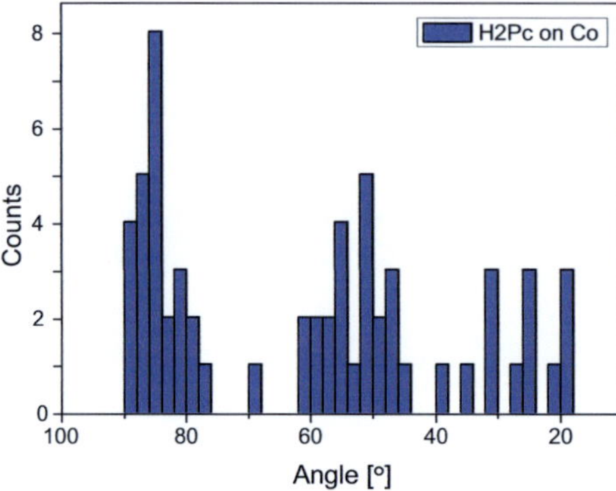

Figure 4.2: Distribution of the adsorption angles for H_2Pc on Co. There are two maxima located at $-85°$ and $-55°$. Additional scatter is visible around $-25°$. The angle distribution with a difference of $30°$ between the maxima is consistent with a fourfold molecule adsorbed on a sixfold surface.

itself account for the electron transmission between the two interfaces. The latter influence is studied by using molecules with distinct central ions. By this, the influence of the central cation on the molecular orbitals and on the transport is studied. Thus, the three different types of Pc molecules were examined on the same substrate, *i.e.* the bare Cu. The current distance curves were measured as described in section 3.4. In order to reduce the influence of noise effects as well as statistical spread, the measurements were performed manifoldly. The obtained histograms are used to calculate the average molecular conductance G_{mol} for the different molecules.

4.2.1 Metal free phthalocyanines on the Cu substrate

The first experiments were performed on the metal free H_2Pc. The absence of a metallic cation in the central cavity reduces the influences on the transport thus being a good starting point. Figure 4.3 (a) shows a typical current distance curve measured on top of an H_2Pc molecule. The first approach displays an exponential increase due to reducing the tunneling barrier width. This is expected since the molecule still lies flat on the surface and thus the system is in the tunneling regime. Below a certain distance, the curve shows a sharp increase in the conductance. This discontinuity indicates the transition from the tunneling to the contact regime and can be explained by a conformational change in the molecular geometry: the molecule is partly lifted from the surface and forming a molecular junction between tip and surface. After the transition to the contact regime the slope of the curve is strongly reduced compared to that of the tunneling regime, indicating that the molecular conductance dominates over the remaining tunneling contribution.

After the first approach the tip was kept at the same spatial position on top of the molecule and was retracted to its initial position. Subsequently, the procedure was repeated at the same position $10-20$ times. For the following current distance curves, the behavior was in the most cases different from the first approach. The curve did not start with an exponential increase from low conductances but at a nearly constant and high value. This can be explained by the molecule remaining in the contact regime when the tip was retracted. This indicates a strong binding between the W tip and the molecule. For these curves, there existed no value for the conductance in the tunneling regime to calculate the molecular conductance. Thus, G_{tun} of the first approach was used for the subsequent data as well. However, there were few measurements where the contact was broken when the tip is retracted.

After the last approach the tip was retraced not only to the initial position but so far that the set-point current of the feed-back loop was reached. During the retraction the molecule still remained attached to the tip. Since in the contact regime, the conductance was higher, the tip had to be retracted so far that the connection was finally broken. In the topographical image (see inset of Fig. 4.3 (b)), this can be seen

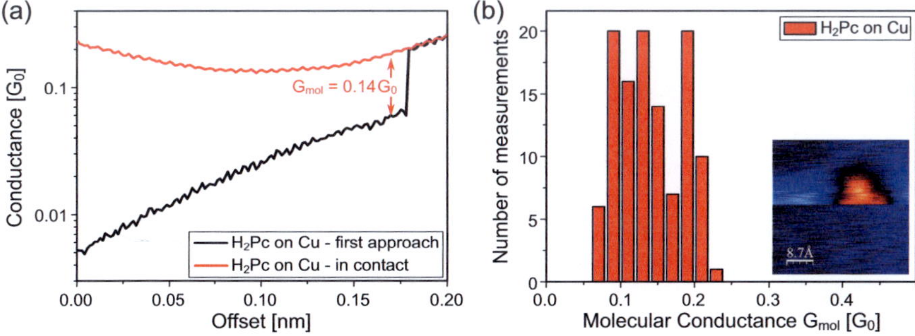

Figure 4.3: (a) Typical distance curve measured for H_2Pc on Cu ($V_{\text{tun}} \approx 10\,\text{mV}$, $T = 4.3\,\text{K}$): exponential increase in the tunneling regime followed by a jump to the contact regime in the first approach (black curve). Then the molecule remains in the contact regime, shown in red averaged over nine individual curves. (b) Histogram of 124 individual measurements of the molecular conductance. The inset displays the topographic scan during obtaining the distance curve, showing the ripping off of the molecule.

by the molecule being ripped off the surface during the retraction and remains attached to the tip. This stresses the stronger binding of the molecule to the less noble W tip than to to Cu substrate.

For H_2Pc on Cu, 124 curves were measured on different molecules and shown in the histogram in Figure 4.3 (b). The distribution shows an approximately Gaussian behavior as expected for the statistical spread. From this distribution the average molecular conductance can be calculated:

$$G_{H_2Pc}^{Cu} = (0.176 \pm 0.012)\,G_0.$$ [4.1]

This relatively high value of the conductance is quite impressive, since phthalocyanine is known to be semiconducting in bulk and thin films [109]. Due to the high conductance the power input to the molecule increases strongly with increasing the voltage. This explains the distance curves could only be measured at low tunneling voltages of around 10 mV.

In order to understand these findings the effects influence the molecular transport have to be examined: on the one hand the hybridization between the molecule and the surface strongly influences the electron injection rates while the orbital structure of the molecule is important for the transmission from one interface to the other. Since the binding and thus the hybridization to the noble Cu surface is rather low, it can be explained by a good transmittance due to the delocalized π orbitals of the molecule.

4.2.2 CoPc and MnPc on the Cu substrate

The curves acquired for Pc molecules, where a metal cation is substituted into the central cavity, showed qualitatively a similar behavior as for the metal free H_2Pc. Both MePc molecules showed the same formation of the contact when the tip was approached the first time (see Fig. 4.4 (a)) as well as the tendency to establish a stable contact, which remained intact when the tip was retracted. Since the hybridization between molecule and substrate is responsible for the strength of the bonding, both seem to be dominated by the organic part of the molecule.

However, already the example curves shown in Figure 4.4 (a) suggest that there is an influence of the central atom on the transport. A detailed analysis of the

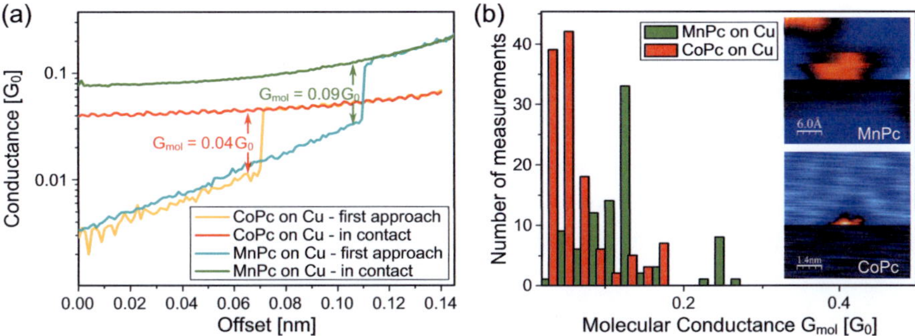

Figure 4.4: (a) Distance curves similar to the H_2Pc case for CoPc and MnPc ($V_{tun} \approx 10\,mV$, $T = 4.3\,K$). (b) The histograms of both CoPc and MnPc are shifted to lower conductances [CoPc: $(0.063 \pm 0.004)\,G_0$, MnPc: $(0.123 \pm 0.006)\,G_0$]. In both cases the molecule is ripped off the surface, seen in the topographical image.

measured conductance values is shown in the histograms of Figure 4.4 (b). The difference between the two molecules can be directly seen. The average conductances are calculated to:

$$G_{CoPc}^{Cu} = (0.063 \pm 0.004)\,G_0 \qquad [4.2]$$

and

$$G_{MnPc}^{Cu} = (0.123 \pm 0.006)\,G_0. \qquad [4.3]$$

4.2.3 Changing the molecular orbitals: the influence of the cation

The conductances for the MePc are reduced when compared with the one of the metal free phthalocyanine $(0.176\,G_0)$. The conductances decrease from H_2Pc to MnPc to CoPc. The strength of the bonding between the Cu surface and the different molecules showed a similar behavior. This suggest that the hybridization of the orbitals of the molecules is independent of the central atom. Thus, the hybridization cannot be responsible for the big change in the transport.

Instead the metal cations must influence the molecular orbitals in such a way that the transport is reduced compared to the H_2Pc. In section 2.2.2 the elastic tunneling spectroscopy was introduced as technique to measure the local density of states. It is one of the big advantages of the STM to allow an extensive study of the molecules when they are already attached to one of the electrodes. Elastic tunneling spectroscopy was used to image the LDOS for the distinct molecules on the Cu substrate and to study the differences in their molecular orbitals. Figure 4.5 show the normalized dI/dV curves for H_2Pc and CoPc.

The green curve shows the LDOS of H_2Pc. Different peaks are visible and can be connected to the molecular orbitals. The first peak below ε_F ($-370\,$meV) can be associated to the HOMO of the H_2Pc an Cu(111), the first one above ε_F ($+390\,$meV) with the LUMO. This results in a HOMO-LUMO gap of 760 meV. Additional orbitals are observed at $\sim -0.8\,$eV (HOMO^{-1}) and $\sim 1.2\,$eV (LUMO^{+1}). For the transport, the orbitals close to the Fermi energy, $i.e.$ the HOMO and LUMO, are the most important ones. Both maxima are located away from the Fermi edge, representing a semiconducting molecule. However, the overlap with ε_F due to broadening of the molecular orbitals results in the observed high conductance value of H_2Pc. The broadening is caused by thermal effects as well as the hybridization of the molecule with the surface.

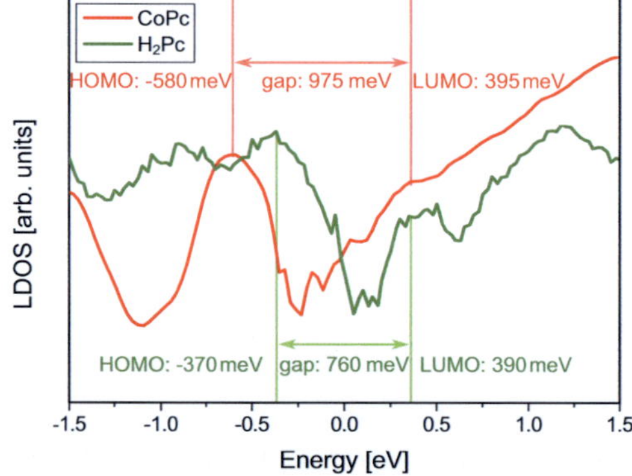

Figure 4.5: LDOS averaged over the sidegroups of the molecules, calculated by normalization of the dI/dV taken at $T = 4.3\,$K with I/V. The different molecular orbitals are visible as peaks in the normalized dI/dV. For H_2Pc the HOMO (-370 meV) and LUMO (390 meV) are located approximately the same distance below and above ε_F, respectively. So the molecule has an HOMO-LUMO gap of 760 meV. For CoPc the LUMO (395 meV) is still in the same energy range as for H_2Pc. However, the HOMO is shifted by approximately 200 meV to $-580\,$meV resulting in an HOMO-LUMO gap of 975 meV. This can explain the reduced conductance for CoPc on Cu.

Similar observation can be made for the CoPc. The LUMO can be connected to the peak at 395 meV and thus is approximately at the same energy as the LUMO of H_2Pc. The HOMO of CoPc is located at -580 meV. In this case an energy shift of the level compared to H_2Pc is observed. The HOMO is shifted ~ 400 meV further away from ε_F. This reduces the overlap of the HOMO with the Fermi level and thus its contribution to the transport. The insertion of the Co ion into the molecule causes a shift of the HOMO resulting in an increase of the HOMO-LUMO gap. This can explain the reduced conductance of CoPc compared to H_2Pc. However, for a complete understanding theoretical calculation for the molecules in contact with the surface would be helpful.

The difference between CoPc and MnPc can be explained in the same way. Theoretical calculation for the molecules in vacuum showed that the HOMO-LUMO gap is reduced for MnPc [122] compared to CoPc [123]. This already indicates the reduced conductance of the CoPc. However, these calculations are performed for the molecule isolated in the gas phase. As the next section will show the surface plays an important role on the electron transport. Thus, a theoretical calculation for the molecules in contact would be helpful. As stated in [122] this is quite complicated for the MnPc since spin effects strongly influences the formation of the metal-organic compound.

4.3 Influence of the surface on the transport: hybridization effects

In order to study the hybridization effects and thus their influence on the transport the electrode material was changed from Cu to Co. The results on the noble Cu surface showed a weak bonding to the surface and therefore a weak hybridization. By changing to the less noble Co islands the transport should be strongly influenced due to a distinct hybridization: a stronger hybridization always causes a higher probability to inject electrons from the source electrode into the molecule or extract it from the molecule to the drain electrode, respectively. To understand how the transport is exactly influenced by the electrode material the same experiments as before were carried out on the Co islands and compared with the results on bare Cu. Subsequently, to get a better understanding of the hybridization effect, elastic tunneling spectroscopy was applied and compared for the different substrates.

4.3.1 H_2Pc and CoPc on Co islands

In the beginning, the curves measured on Co show a similar behavior as on Cu. The curves display an exponential behavior in the tunneling regime followed by a transition to the contact regime indicated by the discontinuity in the conductance. The first observed deviation was that molecular contact was broken when

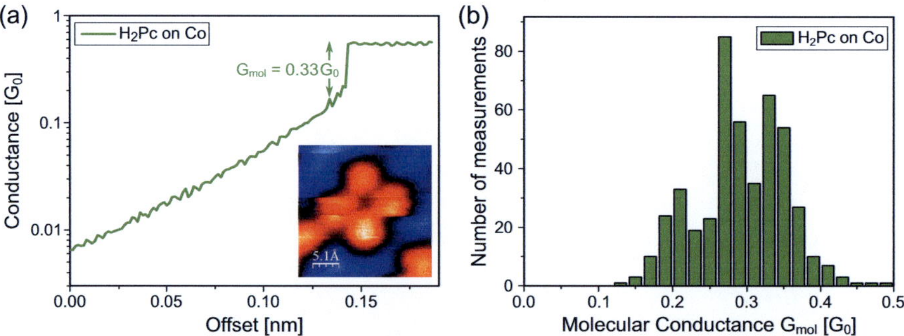

Figure 4.6: (a) Typical distance curve measured on H_2Pc on Co ($V_{tun} \approx 10\,$mV, $T = 4.3\,$K). Here the molecules flip back after every approach. The inset stresses the higher bonding to the substrate with the molecule remaining on the surface. (b) Histogram for H_2Pc on Co (465 curves). The distribution is distinctly shifted to higher conductances compared to Cu.

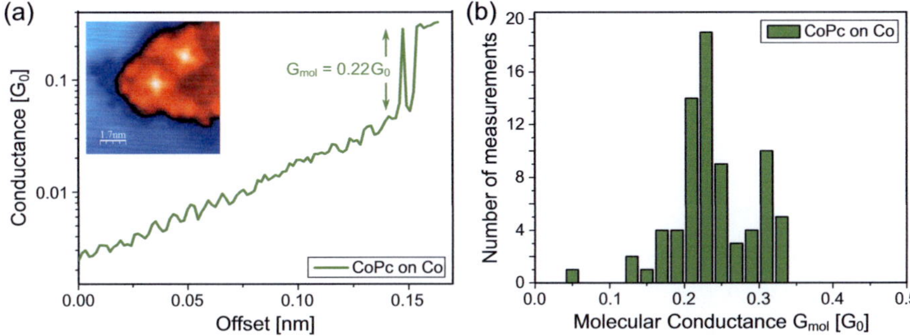

Figure 4.7: (a) Exemplary distance curve for CoPc on Co ($V_{\text{tun}} \approx 10\,\text{mV}$, $T = 4.3\,\text{K}$). A stronger bonding is indicated in the inset by the remaining of the molecule on the surface. (b) The histogram (76 curves) is shifted to higher values compared to the measurements on Cu.

the tip was retracted to its initial position. Thus, the distance curve recorded subsequently started again in the tunneling regime. This indicates a higher bonding of the molecule to the less noble Co surface compared to that of bare Cu. This finding is confirmed by the molecule remaining on the surface when the topographical scan is resumed (see inset of Fig. 4.6 (a)).

The detailed analysis of the conductance distribution reveals also a strong influence of the hybridization on the transport. The histogram is shifted to higher conductances (see Fig. 4.6 (b)). For H$_2$Pc the average molecular conductance can be extracted to be

$$G_{\text{H}_2\text{Pc}}^{\text{Co}} = (0.296 \pm 0.003)\,G_0, \qquad [4.4]$$

which is nearly a factor of two higher than on Cu. Since the substituted surface material is the only difference to the previous experiment, this increase of the conductance can be explained by a stronger hybridization between the molecule and the Co(111) surface and thus a higher electron injection rate.

Similar observations hold for the CoPc molecules on Co (see Fig. 4.7). Again the bonding to the surface is stronger compared to the Cu case indicated by the transition back to the tunneling regime after every approach and the remaining of the molecules on the surface when the tip was retracted. Thus, the stronger hybridization is not only present for the metal free phthalocyanine. Hence, the distribution of the individual molecular conductances is also shifted to higher values. The average conductance of

$$G_{CoPc}^{Co} = (0.239 \pm 0.006)\,G_0 \qquad [4.5]$$

is increased by a factor of approximately four compared to the case of the bare Cu indicating the strong influence. However, the conductance of CoPc is still smaller than that of H_2Pc. The absolute difference between H_2Pc and CoPc remains nearly constant for both substrates indicating that the central atom only slightly influences the hybridization while the changes in the molecular orbitals still cause reduction of the transmission.

4.3.2 The LDOS on the different substrates

The molecular orbitals are not only influenced by the central metal cation. Instead, the distinct hybridization with the different substrates should show an even higher effect. Thus, spectra of the two molecules on Cu and Co were taken to gain insight into the hybridization effects. In Figure 4.8 (a) the results for the normalized dI/dV

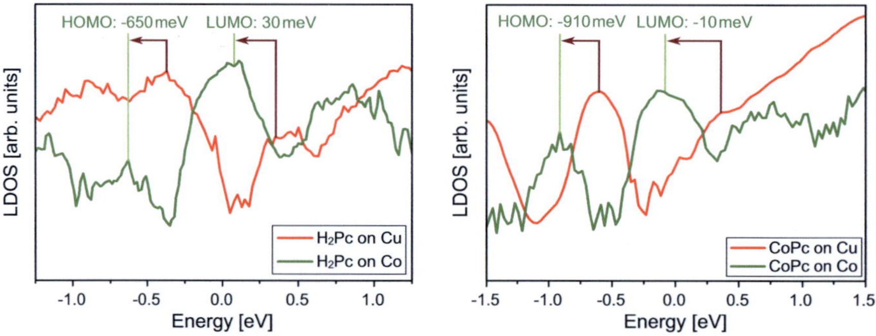

Figure 4.8: The LDOS measurements for H_2Pc and CoPc on Cu from Figure 4.5 are compared with the same measurements obtained at Co ($T = 4.3\,K$). (a) For H_2Pc the HOMO ($-650\,meV$) and LUMO ($30\,meV$) are shifted to lower energies with respect to the previous measurements on Cu (HOMO: $-370\,meV$; LUMO $390\,meV$). This results in the LUMO being located at the Fermi edge and thus can explain the strongly enhanced conductance which is observed for Co. (b) The same observations hold for CoPc. The HOMO ($-910\,meV$) and LUMO ($-10\,meV$) are shifted to lower energies as well, resulting in the latter being located at ε_F.

curves proportional to the LDOS are depicted for H_2Pc. The previous results on Cu are compared to Co. For H_2Pc on Cu the peaks displaying the HOMO ($-370\,meV$) and LUMO ($+390\,meV$) on Cu are shifted to lower energies while the overall shape is conserved. The new HOMO ($-650\,meV$) is located much further below ε_F and thus contributing less to the transport. The electron transport on Co is mainly dominated by the LUMO. Since it is located closely above the Fermi level ($30\,meV$), it can be easily accessed by the electrons from the electrodes. In contrast to the semiconducting level structure for H_2Pc on Cu, the molecules becomes metallic on Co. Theoretical DFT calculations have been performed for the H_2Pc molecules. A comparison with them shows that the shift is due to an electron transfer between the N atoms and the surface. The LUMO which has a significant contribution of the nitrogen is shifted towards the Fermi edge. This shift of $\sim -350\,meV$ results in the LUMO being directly located at the Fermi edge. This hybridization effects also reflect a fundamentally different adsorption mechanism: physisorption versus chemisorption.

The same observation can be made for CoPc on Co. The molecular orbitals are shifted to lower energies compared to Cu (see Fig. 4.8). The shift is of the same size ($\sim 400\,meV$) as for H_2Pc. Again, this results in the LUMO on Co ($-10\,meV$) being located close the Fermi edge. The HOMO is dislocated to $-910\,meV$. For CoPc strong shift of the HOMO to lower energies compared to H_2Pc is observed due to the insertion of the central metal ion, confirming the results on Cu. This is another proof for the weak influence of the central cation on the hybridization.

In Figure 4.9 spatial resolved dI/dV maps of CoPc are shown for different energies on the two substrates. The shape of the molecular orbitals strongly depend on the energy and the surface material. On Co islands, the signal of the central Co ion of the molecule dominates the images and thus masking the low LDOS of the organic sidegroups.

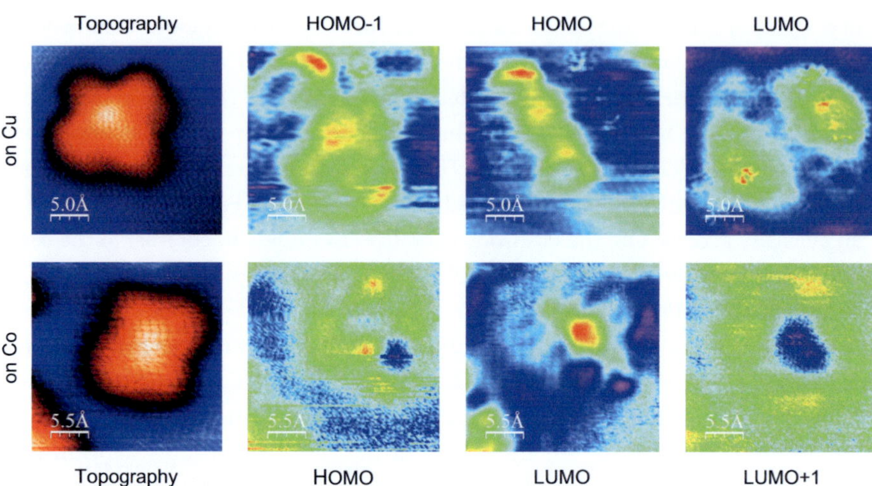

Figure 4.9: Spatial resolved images of the molecular orbitals of CoPc on the different surfaces. The images on Cu are taken at $V = -1.54\,\text{V}$ (HOMO^{-1}), $V = -0.56\,\text{V}$ (HOMO), and $V = 0.38\,\text{V}$ (LUMO), the ones on CO at $V = -0.95\,\text{V}$ (HOMO), $V = -0.04\,\text{V}$ (LUMO), and $V = 0.90\,\text{V}$ (LUMO^{+1}) ($T = 4.3\,\text{K}$). The molecular orbitals on Co are masked by the strong signal of the central Co cation.

4.4 The process of forming the contact: soft vibron excitation

The extensive studies on assembling CoPc and H$_2$Pc molecular junctions indicate that the molecular jump occurs more likely when the tip approaches the isoindol sidegroups rather than the center of the molecule. To understand this observation, ISTS measurements were carried out by measuring the second derivative of the tunneling current, which can reveal vibronic excitations created by tunneling electrons [21]. Indeed, when the kinetic energy of the tunneling electrons is sufficient to create an inelastic excitation in form of a vibron, a pair of peak and dip in the d^2I/dV^2 appears [124]. Spatial resolution was achieved by recording inelastic spectra at different positions of the Pc molecule. Figure 4.10 (a) shows the d^2I/dV^2 spectrum of a CoPc molecule on Co/Cu(111). An inelastic excitation of about 20 meV appears in the spectrum as an antisymmetric combination of a peak and a dip at ± 20 meV, respectively. This excitation shares the same energy range as the excitation energy of benzene-substrate vibrons [125]. To locate the vibronic excitation within the molecule, the values of the inelastic spectrum at $\pm 20\,$mV is plotted as a function of tip position atop the molecule (see insets of Fig. 4.10) (a). A high inelastic excitation probability is indicated by a combined dark intensity at negative and a bright intensity at positive bias. To stress this finding, the two images were subtracted — shown in the inset of Figure 4.10 (b) — resulting in four bright peaks located on the sidegroups of the molecule. Thus, one can see that the excitations are located on the sidegroups of CoPc and can therefore be associated with a vibrational mode of the molecular sidegroup.

The same vibron could be observed in $I(V)$ curves measured in contact (shown for CoPc on Co in Fig. 4.10 (b)). To acquire these curves the spectroscopic technique was combined with the approach technique. Before recording the spectra, a certain voltage was applied to the z piezo. In this way the tip was approached toward the surface and kept in the proximity of the molecule. Thus, there was a certain possibility of the transition of the molecule from the tunneling to the contact regime. The transition can be seen in the $I(V)$ curve like for the distance curves as a discontinuity (see Fig. 4.10 (b)). After forming the contact, the $I(V)$ curve is completely

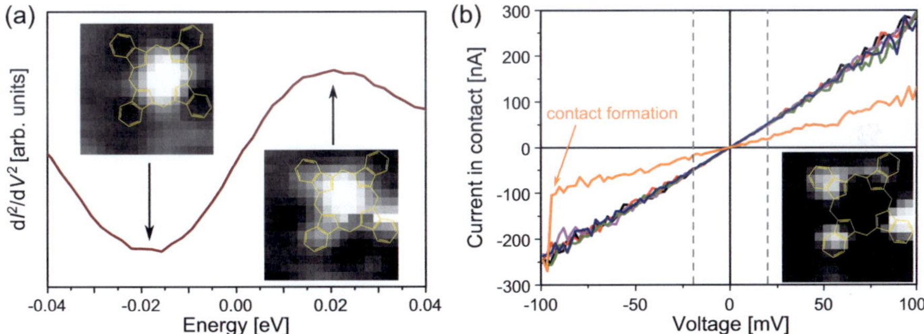

Figure 4.10: (a) $\mathrm{d}^2I/\mathrm{d}V^2$ curve for CoPc averaged over the side group of the molecule taken at $T = 4.3\,$K. One observes a excitation located at around $20\,$meV. The insets show the $\mathrm{d}I/\mathrm{d}V$ taken at $\pm 20\,$meV. The dip-peak structure is clearly constrained to the side groups of the molecule, visualized in the inset of (b), showing the difference of the two images. (b) $I(V)$ curves measured with applying voltage on the z piezo. The orange curve indicates the contact formation. Subsequent measurements were taken in the contact regime, when the molecules bridges the tunneling gap. At $20\,$mV the noise of the curve strongly increases, indicating that the soft vibron of the molecule is excited.

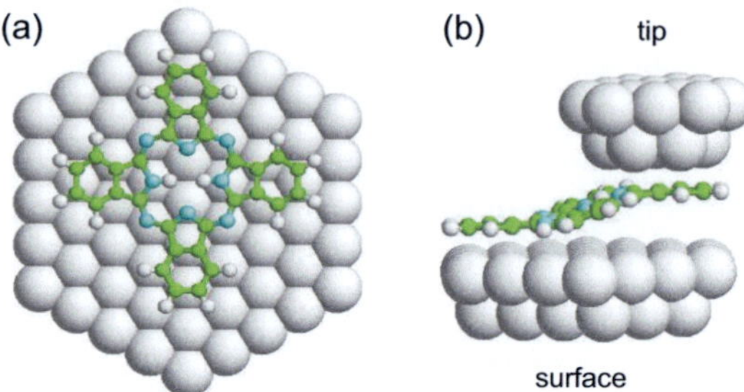

Figure 4.11: Theoretical geometry optimization: (a) flat and (b) contact position. The H_2Pc molecule is adsorbed in bridge position on the surface. This means that the two central hydrogens are aligned on two neighboring hollow sites. When the tip is close to the surface, there is a second local minimum for the alignment of the molecule where the molecule forms a contact between surface and tip.

linear in the voltage range which can be applied without destroying the molecular contact. This indicates a metallic behavior in the case of the molecule adsorbed on Co which is in accordance with the high conductance values and the LDOS shown in Figure 4.8.

For low energies, the noise of the curves is very low, while it sharply increases when the vibron excitation energy of approximately 20 meV is exceeded. This points out the possibility of the molecule being vibronicly excited in the contact regime as well. For the reduction of noise effects the conductance measurements have to be performed below 20 meV.

Additionally, theoretical calculation have been performed for the geometry optimization of the molecule inside the junction. For the tip being close to the surface, there are to possibilities how the molecule can be aligned: either in flat geometry representing the tunneling regime or in the contact geometry (see Fig. 4.11). In the energy landscape these two geometries correspond to two local minima separated by a small energy barrier. The transition between these two states is expected to follow the lowest energy path between the two states. Thus, the transition is related to the lowest lying vibronic excitation which is the reason for forming the contact but also for breaking it.

5 Giant magnetoresistance of single phthalocyanine molecules

For the application in data storage devices the most interesting point is to realize molecular junctions exhibiting a large magnetoresistance. In the previous chapter the method of contacting single molecules was introduced using the example of different phthalocyanine molecules. The results showed a strong dependence on the metal ion forming the compound with the organic part as well as on the substrate. In order to reduce the different influences the GMR experiments will be carried out for H_2Pc junctions since the impact of the central atom is not yet completely understood. They have the additional advantage, that they also exhibit a higher conductance. The lower areal resistance is essential for further high frequency application needed in data storage.

The previous results showed that the Co islands are suited for contact measurements. Since they offer the requested magnetic properties (see section 3.2), they are used as one of the electrode. The tip material of choice is the Co coated W tip. This ensures an out-of-plane sensitivity and a symmetric tunnel junction. In this chapter the first successful measurement of the magnetoresistance of single molecules in the ballistic regime will be reported. The GMR ratio of the $Co/H_2Pc/Co$ junction exceeds the according TMR ratio by a magnitude. This is a promising result for possible future applications.

Theoretical calculations were performed in order to explain the huge effect of the Pc molecules on the magnetotransport. This allows to connect the effect with the hybridization of the molecule with the Co of tip and substrate.

For future applications even higher GMR ratio would further improve the usability. Since the phthalocyanine showed a large effect on the magneto resistance and their practicability in the contact measurements was proven, the idea was to change the electrode material. Thus, experiments on a second system, the $Mn/H_2Pc/Fe$ junctions, were carried out, since Mn should show a higher spin polarization around the Fermi edge compared to Co.

5.1 The magnetoresistance of Co/H$_2$Pc/Co junctions: the first single molecule GMR

The experiments were performed on identical samples as the non-magnetic measurements. In order to get out-of-plane sensitivity as well as a symmetric junction, a Co-coated W-tip (10 monolayers) was used. Topographical measurements show the same behavior as before. Nevertheless, the samples had to be characterized magnetically at first. Spin polarized measurements of the differential conductance (dI/dV) carried out on the bare Co islands (see Fig. 5.1 (b)) reproduced the results of previous works [76]. The optimistic TMR ratio

$$\mathrm{TMR} = \frac{G_> - G_<}{G_<} \qquad [5.1]$$

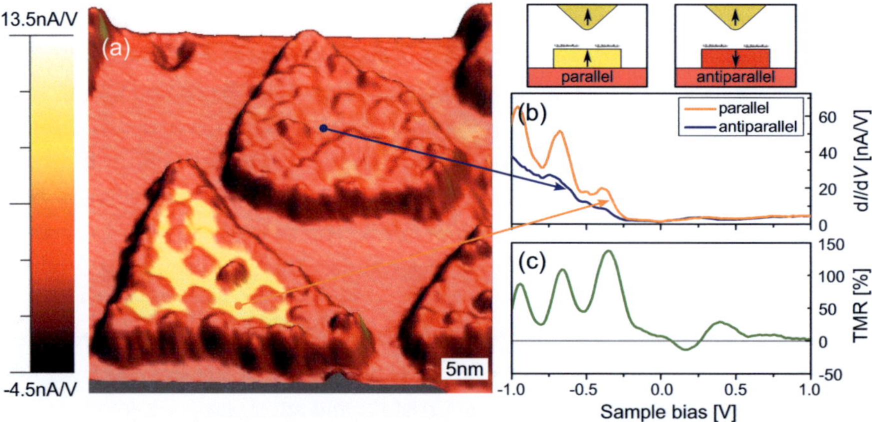

Figure 5.1: a) Topographic image of H$_2$Pc molecules adsorbed onto two Co islands on the Cu(111) surface ($T = 4.3$ K). The color code displays the measured dI/dV map at -310 mV. One can distinguish between the two islands species with a magnetization parallel (in yellow) and antiparallel (in red) with respect to the tip magnetization. b) dI/dV spectra taken on two islands of P and AP type, which clearly reveal spin-polarized states below the Fermi edge. c) Optimistic TMR ratio calculated from the dI/dV spectra. The highest value is measured at around -350 meV, and is used to distinguish between the two islands species. Please note that the TMR ratio around the Fermi edge is of $\sim 5\%$.

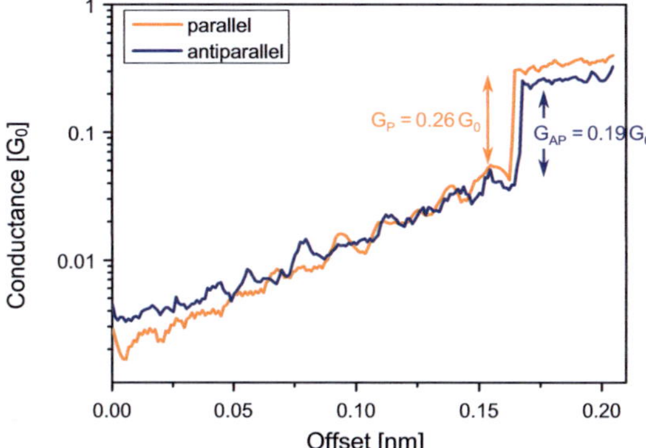

Figure 5.2: Exemplary current distance curves measured on contrariwise magnetized islands ($V_{tun} \approx 10\,$mV, $T = 4.3\,$K). An indication to a dependence of the transport on the relative alignment of the electrodes can already be seen in this curves The same method as in the non magnetic case is used to extract the information on the molecular transport.

calculated from the individual curves (see Fig. 5.1 (c)) show a strong peak at around $-350\,$meV which can be connected to the Co surface state [101]. The variation in the differential conductance at this energy can be used to determine the orientation of the magnetization of the individual islands relative to that of the tip (parallel or antiparallel). Thus, the different curves can also be connected to parallel or antiparallel alignment, respectively. The difference at $\sim -350\,$mV allowed to detect the local magnetization direction of Co islands by recording maps of the local differential conductance at this bias voltage. In Figure 5.1 (a) a combined image of the STM topography and a spatial resolved dI/dV map at $-310\,$mV of the sample is presented. In this image, the topography serves as the actual 3d shape while the colors of the image represent the differential conductance. The contrast in the dI/dV map can directly be connected to the distinct magnetization of the islands, since both islands are pointing upwards and therefore stacking effects can be excluded. Thus, the islands which appear yellow and red can be connected to parallel and antiparallel alignment, respectively.

The differential tunneling magnetoresistance depends strongly on the energy as depicted in Figure 5.1 (c) and shown previously by others [76]. Remarkably, the differential TMR for this Co/vacuum/Co junction is only $\sim 5\%$ at low bias voltage

($< 10\,$meV). This is important since the transport measurements on H_2Pc are restricted to this energy range due to the high risk of destroying the molecules when they are contacted at higher voltages.

For the actual transport measurements the same method as in the non-magnetic studies was applied. To determine the magnetoresistance ratio of the H_2Pc molecule the transport measurements had to be performed on differently magnetized islands. In Figure 5.2, exemplary curves for both magnetization directions (parallel and antiparallel) are shown. The curves reveal the same exponential increase of the conductance reflecting the decreasing tunneling barrier width, followed by the transition to the contact regime, visible as a discontinuity in the distance curves (compare Fig. 4.6). Identical tip conditions were ensured by recording the traces from the two islands during the same scans. This approach eliminated magnetostriction of the two ferromagnetic electrodes, as neither the tip nor the island magnetization was switched by an external magnetic field.

Indeed, already the sample curves shown in Figure 5.2 display a marked dependence of the conductance on the island's spin polarization. In the particular measurement, values of $G_P = 0.26\,G_0$ and $G_{AP} = 0.19\,G_0$ were found. These values are in the same range as for the measurements with the non-magnetic W-tip [4.4]. Thus, the hybridization of the molecule with Co and W, respectively, are of the same order since both are base metals.

These high values of the conductance are in sharp contrast to the exponentially low values that have been reported for almost insulating molecular TMR layers [38,126] and single molecular devices based on C_{60} [127]. This is a big advantage for a future application in hard disks, since a too high areal resistance of these devices would inhibit high frequency applications.

To eliminate the influence of the statistical spread and of the noise, each such measurement was repeated several hundred times on about 10 different molecules on the two islands and the distribution of the conductances in parallel and antiparallel was evaluated (see Fig. 5.3). The width of the resulting conductance distribution is relatively narrow, in particular not logarithmic, when compared to previous work [18]. The broadening and possible structural elements reflect the variation in contact and binding geometries, but also fluctuations in the spin-polarized density of states on the islands [76, 128]. Gaussian fits were used to determine the average conductances and the GMR from the measurement statistics. The average conductances

$$G_P^{Co} = (0.253 \pm 0.005)\,G_0 \qquad [5.2]$$

and

$$G_{AP}^{Co} = (0.158 \pm 0.005)\,G_0 \qquad [5.3]$$

were found. These values result in an optimistic GMR ratio of

$$\text{GMR}_{Co} = \frac{G_P^{Co} - G_{AP}^{Co}}{G_{AP}^{Co}} = (61 \pm 9)\,\%. \qquad [5.4]$$

Surprisingly, the GMR ratio obtained at 10 mV is one order of magnitude larger than the differential TMR found for direct tunneling between the magnetic tip and the bare Co surface.

Theoretical description

Although the molecule itself is non-magnetic, the influence on the spin transport revealed to be very high. The mechanism of the GMR already implies that hybridization effects play an important role. For a better understanding of the pro-

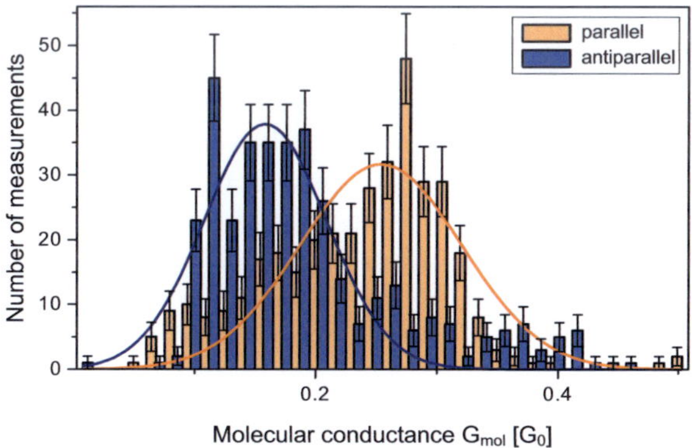

Figure 5.3: Distribution of the measured molecular conductances (381 times parallel/ 366 times antiparallel). The difference between the two channels can be clearly seen. The Gaussian fits are used to determine the two different conductances and the GMR ratio ($(61 \pm 9)\,\%$).

cesses responsible for the large GMR ratio, transport calculations based on density functional theory (DFT) employing the non-equilibrium Green's function (NEGF) formalism and the TURBOMOLE package [129] were performed. Spin-polarized transport properties were calculated in the linear response at low bias voltage for the two junction geometries. The conductances in the two different regimes were calculated individually and are represented by two independent curves (see Fig. 5.4). On a qualitative level these calculations reproduce the experimental results very well: they reflect the exponentially increasing conductance in the tunneling regime, $G_{\text{tun}}(d) \approx e^{-\beta d}$, when the distance d between the two electrodes is still large. The slope is independent of the relative alignment of the electrode magnetizations. Further, the computational value of the slope $\beta^{\text{theo}} = 1.87\,\text{Å}^{-1}$ — connected to the work function $W^{\text{theo}} = 3.24\,\text{eV}$ — is in good agreement with the experimental value ($\beta^{\text{exp}} = (1.9 \pm 0.3)\,\text{Å}^{-1}$; $W^{\text{exp}} = 3.2\,\text{eV}$).

Reproducing the experimental findings, the variation of $G_{\text{cont}}(d)$ with the contact distance d is very weak once the contact is established. However, it is still sensitive to the relative orientation of the magnetization of the electrodes. $G_{\text{AP}}(d)$ is always much lower than $G_{\text{P}}(d)$. While the experimental conditions themselves cause a transition from the tunneling to the contact regime, this process is not intrinsically included in the theory. Thus, the point to calculate G_{mol} have to be chosen manually. For a quantitative comparison with the experiment, the GMR_{theo} ratio is considered at the distance d at which the ratio $r = G_{\text{tun}}/G_{\text{cont}}$ matches the value $r \approx 4$ found experimentally. Thus, $\text{GMR}_{\text{theo}} \approx 65\,\%$ is only weakly dependent on d. This is already in very good agreement with the experimental results. However, a slightly higher value is expected for the theoretical value, since in the calculations the relative magnetization of tip and sample is completely parallel while this cannot be ensured in the experiment. In the case of the experiment, the magnetization of the Co islands is fixed due to the high surface anisotropy. Thus, it is always pointing exactly perpendicular to the surface. The situation for the tip is much more complicated. Although the prepared Co films on the W-tips have an out-of-plane anisotropy, the complex structure of the tip apex can strongly influence the spin polarization of the tip and thus reduce the measured GMR ratio.

The Pc molecules are characterized by an energetically isolated HOMO and a nearly doubly degenerate LUMO [130]. The HOMO levels, corresponding to the aromatic group, hybridize only very weakly with almost no amplitude on the bridging ni-

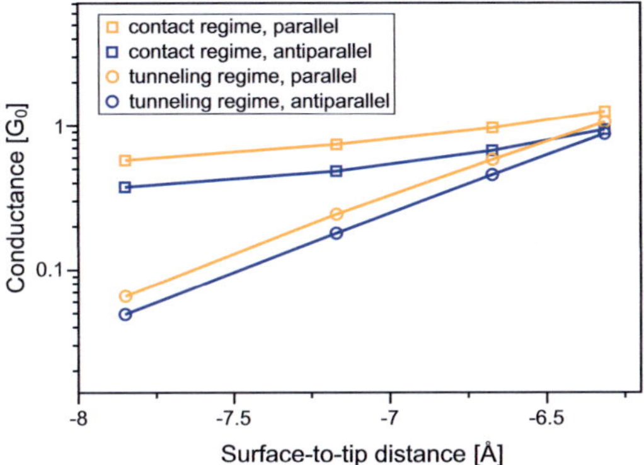

Figure 5.4: Theoretical DFT calculation for the molecular conductance in the two different adsorption and spin geometries. In the calculations the two regimes are separated, while for the experiment the transition between the regimes is indicated by the discontinuity. The theoretical conductances are also extracted by subtraction of the two regimes at comparable distances with the experiment. This results in a theoretical GMR of 65 %, which is in good agreement with the experimental value.

trogen. By contrast, the LUMO states are located on two out of the four aromatic groups, with a strong hybridization to all N atoms forming the inner macro-cycle. Since the N bond to Co includes states at the Fermi energy ε_F (see sect. 4.3.2), transport should occur via the quasi-degenerate LUMO level. This fact is confirmed by examining the transmission per spin direction around the Fermi energy across a junction, magnetized parallel and antiparallel, respectively (see Fig. 5.5). For all configurations, G_P near ε_F is indeed dominated by a peak centered slightly above ε_F that indicates transmission through the LUMO level — this is in accordance with the measured LDOS (see Fig. 4.8). The peak width is determined by the strength of hybridization of the molecular LUMO orbital with the substrate and tip states via the N-Co binding. Since the surface DOS of the Co electrodes at ε_F is enhanced for minority spin electrons as compared to the majority ones, the LUMO broadening is strongly spin dependent. This difference in LUMO broadening for the two spin channels has a direct impact on the GMR measured across the molecular junction and thus causes the high GMR ratio.

For a qualitative understanding of the mechanism, one has to consider the molecular orbitals in real space for the four different spin configurations shown in Figure 5.5. The two orbitals in the parallel case are manifestly distinct. The hybridization of the minority channel of the Co surfaces with the LUMO give rise to an orbital that is completely delocalized from the tip to the surface. The electrons in this channel can move easily from one electrode to the other and contribute strongly to the transport. The weak hybridization of the majority channel with the molecule results in the absence of such an extended state and thus in a low contribution to the transport. Nevertheless, due to the minority channel there is a high conductance for the parallel alignment. If the magnetization of one of the electrodes is

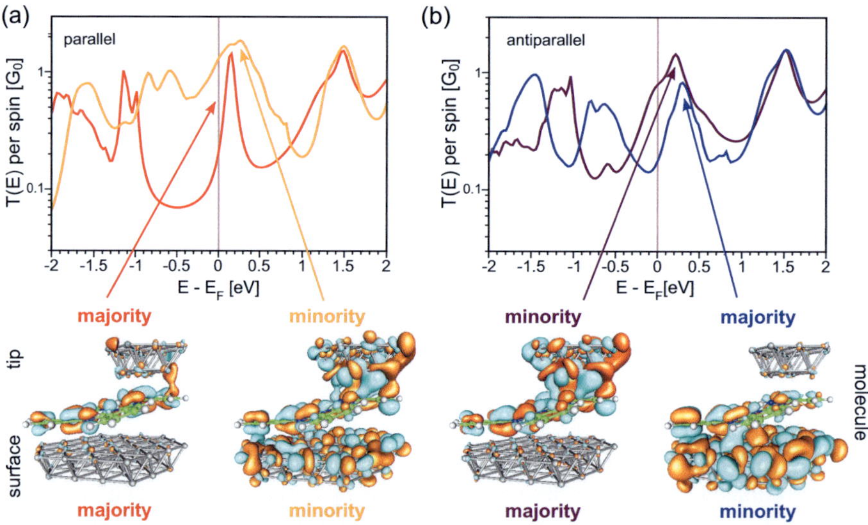

Figure 5.5: Energy dependent transmission for the four channels contributing to the GMR effect: for each alignment (parallel and antiparallel) there are two spin channels. The distinct hybridization of these channel is the reason for the high GMR ratio. For all transmissions there is a peak close above ε_F. The difference is the peak width which is due to the different hybridization. The minority channel for the parallel alignment contributes most to the transport since it is the broadest and thus strongest at the Fermi edge. The four lower images visualize the different states and thus the hybridization. In the parallel alignment the minority channel is forming a state extended from tip to surface. This results in a strong contribution to the conductance. For the antiparallel alignment, both channels hybridize either with the tip or the substrate. Thus, the conductance is strongly reduced.

switched the situation changes drastically. In this case, the LUMO hybridizes with either the tip or the sample, since the majority and minority channel are reversed with respect to each other. Thus, there is no state reaching from one electrode to the other. This quenches the transport for both channels and reduces the conductance compared to the parallel case.

5.2 Mn/H$_2$Pc/Fe junctions: optimization of the GMR

In the previous section, the results of the first successful measurement of the GMR of an individual molecule was shown. The GMR ratio of approx. 60 % is already quite high, when compared to the TMR for the same setup of electrodes and the same energies. Nevertheless, even higher GMR ratios would increase the signal-to-noise ratio and therefore improve the performance in sensors and read heads. To get closer to reach this goal the system must be optimized in such a way that the GMR increases to higher values. The Pc molecules showed, that they have a strong influence on the magnetotransport and thus are a prommissing molecular GMR system. Without changing the molecule itself, it remains the possibility to

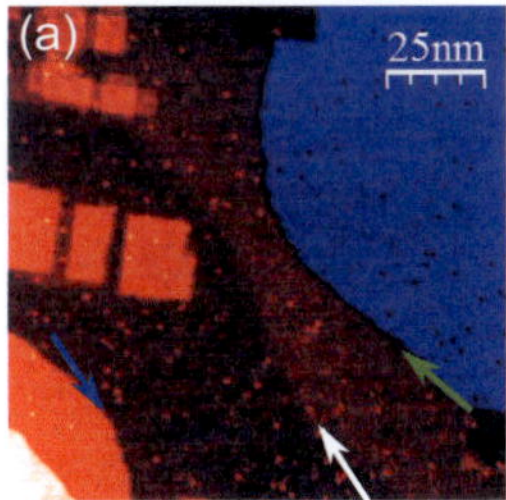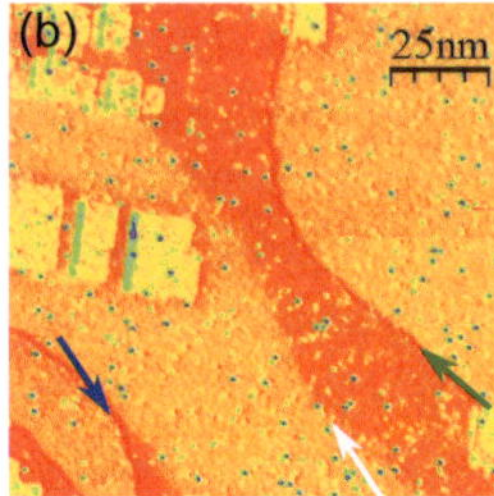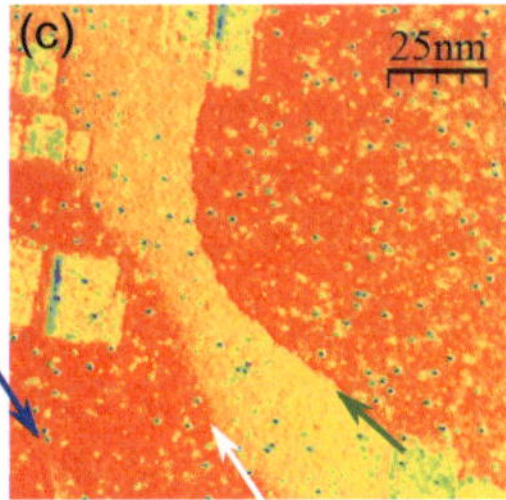

Figure 5.6: (a) Topographical image of Mn grown on an Fe(100) whisker ($V = 199$ mV, $I = 2$ nA, $T = 4.3$ K). (b) dI/dV map taken before reversing the external field ($V = 199$ mV, $T = 4.3$ K). (c) dI/dV map after reversing the external field ($V = 212$ mV, $T = 4.3$ K). The three arrows indicate the main magnetic features of the Mn/Fe(100) surface. Green: a Mn step edge at the surface, that is visible in the topography and the dI/dV maps. White: a buried Fe step edge, that is only slightly visible in the topography but can clearly be seen in the dI/dV map. Blue: coincident Fe and Mn step edges. The magnetic effect is compensated.

change the electrode material. Due to the relatively low spin polarization of the Co DOS around ε_F the GMR is confined to relatively small values. The idea is to use an electrode material with a higher spin polarization of the DOS at ε_F. The well studied Mn/Fe(100) offers this possibility and is thus a good choice for carrying forward the experiments. Since the Mn films show an in-plane anisotropy an in-plane sensitve Fe tip was used for the experiments.

The main properties of the Mn/Fe(100) surface have already been introduced in section 3.2. In Figures 5.6, a topographic image and two $\mathrm{d}I/\mathrm{d}V$ maps after the deposition of H_2Pc on Mn/Fe(100) is presented. In these images the important magnetic features are already visible. A Mn step edge clearly visible in the topography is indicated by the green arrow and causes a reversal of the surface magnetization visible in the $\mathrm{d}I/\mathrm{d}V$ maps. The white arrow points to a buried Fe step edge: in the topography this buried step edge is seen by a slight height difference. In the $\mathrm{d}I/\mathrm{d}V$ maps, the reversal of the magnetization can be seen as clearly as before. The blue arrow alludes to another buried step edge, that partly coincides with an Mn step edge. Thus, the magnetization stays the same as both effects compensate.

By applying an external field aligned in the opposite direction with respect to the magnetization of the Fe whisker, the magnetization of the whisker can be reversed. So far, there have been no studies focusing on what happens with the Mn layers on top when the Fe whisker is switched. Figure 5.6 (c) shows the $\mathrm{d}I/\mathrm{d}V$ of Mn on Fe(100) map after reversing the field. All effects seen in the previous $\mathrm{d}I/\mathrm{d}V$ map are reproduced while the bright and dark areas are swapped. The magnetization of the Mn layers follows the one of the Fe whisker when the latter is reversed.

In Figure 5.6 a high density of adsorbates is visible. These are H_2Pc molecules as well as undesired dirt particles. The latter are mainly deposited together with the H_2Pc molecules and in contrast to the Cu and Co surfaces sticked to the highly reactive Mn surface. Hence, it was much more difficult to detect intact H_2Pc for controlled transport measurements — seen for example in the inset of Fig. 5.7 (a). Nevertheless, there were some molecules, which could be used to measured distance curves, as shown in Figure 5.7 (a). They showed similar properties as before on Cu and Co. Like on Co, the connection of the tip with the molecule was broken when it was retracted. However, a tendency to higher conductances could be observed. This is a sign for even stronger hybridization with the less noble Mn surface. The distribution obtained from 108 single measurements in parallel alignment

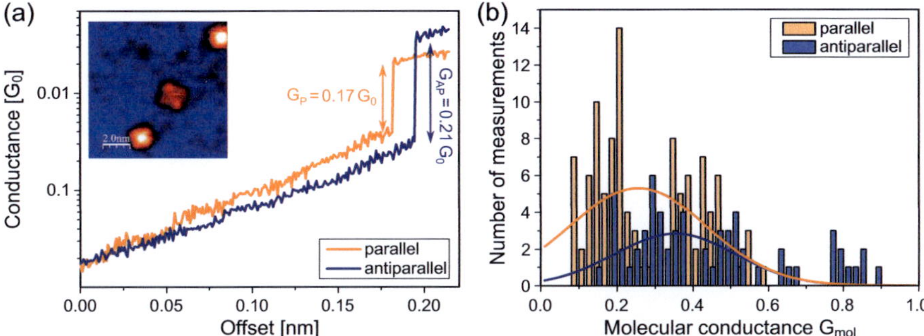

Figure 5.7: (a) Exemplary distance curves measured on H$_2$Pc on Mn/Fe(100) ($V_{tun} \approx 10$ mV, $T = 4.3$ K). Breaking of the contact, when retracting the tip indicates a strong bonding. The conductances for antiparallel spin alignment show a tendency to be higher. (b) The distribution (108 times parallel, 63 times antiparallel) does not show one clear peak for the different alignments. This can be due to different adsorption geometries caused by the stronger binding to the Mn surface. Nevertheless, a tendency to higher conductances for the antiparallel case can be observed. This results in an negative GMR ($-(54 \pm 16)$ %).

and 63 in the antiparallel one is shown in Figure 5.7 (b). Since the measurements on Mn were more complicated than on Co/Cu(111), much less measurements could be performed. Thus, the quality of the distribution is not as good as before. For the parallel case the distribution consists of two individual peaks located at around $0.2\,G_0$ and $0.4\,G_0$. The distribution in the antiparallel case displays even four peaks. There are possible explanations for the substructures in the distributions. On the one hand, the H$_2$Pc molecules sometimes showed a bright center when they were adsorbed on Mn. This could be explained by a different adsorption position an thus distinctly shaped orbitals due to different hybridization mechanisms. The molecules with bright centers showed a tendency to higher conductances. A second observation that was made is that the stronger binding results in a wide spread distribution of the adsorption parameters. This also can cause a higher spread in the conductance.

However, there is a tendency to higher conductances especially for the antiparallel alignment whose conductance even exceeds the conductance for the parallel case.

By extracting the two average conductances one gets:

$$G_{\mathrm{P}}^{\mathrm{Mn}} = (0.286 \pm 0.013)\, G_0 \qquad [5.5]$$

and

$$G_{\mathrm{AP}}^{\mathrm{Mn}} = (0.441 \pm 0.026)\, G_0, \qquad [5.6]$$

respectively. Since $G_{\mathrm{AP}}^{\mathrm{Mn}}$ is bigger than $G_{\mathrm{P}}^{\mathrm{Mn}}$ the optimistic GMR is calculated to:

$$\mathrm{GMR}_{\mathrm{Mn}} = \frac{G_{\mathrm{P}}^{\mathrm{Mn}} - G_{\mathrm{AP}}^{\mathrm{Mn}}}{G_{\mathrm{P}}^{\mathrm{Mn}}} = -(54 \pm 16)\,\%, \qquad [5.7]$$

where the divider is changed to $G_{\mathrm{P}}^{\mathrm{Mn}}$. The negative GMR is a direct consequence of the asymmetric junction geometry. One of the electrodes is formed by Fe while the other consists of Mn. This is the first GMR measurement with a ferromagnetic lead on the one side and an antiferromagnetic one on the other. Since the electrodes consist of different materials, distinct hybridization effects can occur and form a junction with a negative GMR. The hybridization of the molecule with Fe and Mn, respectively, is stronger for the opposite spin channel. Although a new effect, $i.\,e.$ a negative GMR, was observed the GMR ratio could not be increased compared to the $Co/H_2Pc/Co$ junctions. However, for an detailed understanding of the physical properties, theoretical calculations for this system are needed as well.

6 Growth and electronic properties of Chromium acetylacetonate

In the previous chapters, it was shown that the electronic and magnetic junctions based on single molecules exhibit different interesting effects. Especially, the high GMR ratio offers nice possibilities for future applications of molecules in data storage.

But there are further possible approaches how to use metal-organic molecules in the field of data storage. One of them is to employ single molecular magnets (SMM). These molecules are built out of two distinct parts forming a metal-organic compound. The metallic part consists of one or several magnetic ions carrying a spin. The spins of the metal ions is shielded from the environment by the organic ligands. However, the organic part also influences the magnetism of the molecule, in particular its anisotropy, and thus can stabilize the spin. The shielded spins could be used to serve as individual bits of a storage media [131]. As a result, the size of an individual bit is reduced to the size of a molecule which is much smaller than today's bit sizes. For this purpose, such metal-organic compounds have to be put onto surfaces and their magnetic and electronic properties need to be investigated when they are in contact with the substrate. The stability of magnetism in single molecules is often a problem when they are deposited on a surface.

In this chapter, studies on chromium acetylacetonate ($Cr(acac)_3$) consisting of a $Cr(III)$ ion — exhibiting a spin of $1/2$ [25] — shielded by three acetylacetonate ($C_5H_7O_2^+$) sidegroups are presented. In contrast to the Pc molecules, $Cr(acac)_3$ is not a flat molecule. This should result in a better decoupling of the central spin and the surface. Until now, there have been no study on the growth of acac molecules on the surface. Thus, the present chapter will start with the characterization of the growth properties of $Cr(acac)_3$. For the possible application in data storage magnetic and electronic properties are very important. These experiments will be presented in the second part of the chapter.

6.1 Deposition and growth

As described in section 3.3.2 there have been only few STM experiments done with acetylacetonate molecules. So far the studies mainly concentrated on their usage as precursor molecules. As shown by Siddiqi and co-workers [118], Al(acac)$_3$ and Cr(acac)$_3$ are stable under sublimation and thus it is possible to deposit them *in situ* under UHV conditions. The Cr(acac)$_3$ molecule is chosen, since its Cr ion carries a spin of $1/2$ [25]. The molecules were deposited by heating them above their sublimation temperature — as described in section 3.3.2 — on an atomically clean Cu(111) surface. This surface is chosen due to its easy preparation procedure and its well known properties.

Several repeated preparations revealed that the Cr(acac)$_3$ typically did not grow individually on the surface when they were deposited at room temperature. Instead they aligned in chains parallel to the symmetry axes of the surface (see Fig. 6.2 (a)). A close-up scan revealed that the individual molecules were shaped as the letter H (see Fig. 6.2 (d)). This can be explained by the molecule being adsorbed on one of the ligands while the other two form stringers connected by the Cr in the center and thus appear shaped as the letter H — as illustrated in Figure 6.2 (f). Each molecule is connected with the next molecule by one of the stringers aligned in such a way that it lies between the two of the other molecule (see Fig. 6.2 (c)). The next molecule is arbitrarily aligned either on the same side or the opposite side with respect to the molecule before. Thus, they are forming either more straight or more zig-zag lines — as visible in the large scale scan in Figure 6.2 (a)). By these interlinks, the

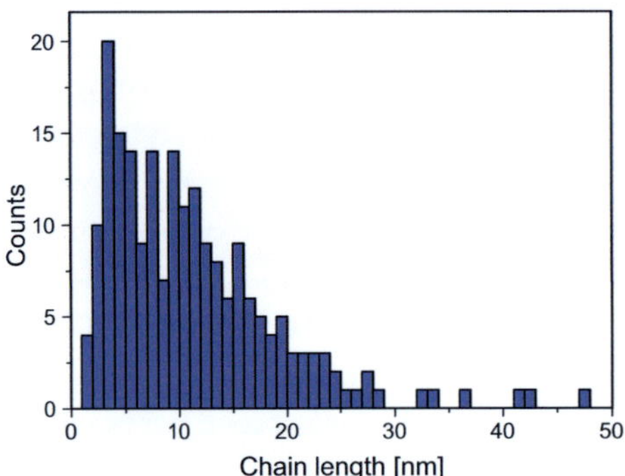

Figure 6.1: Distribution of the chain lengths of Cr(acac)$_3$ on Cu(100). There is a maximum for small chain lengths (2-8 molecules) while there are chains consisting up to 60 molecules.

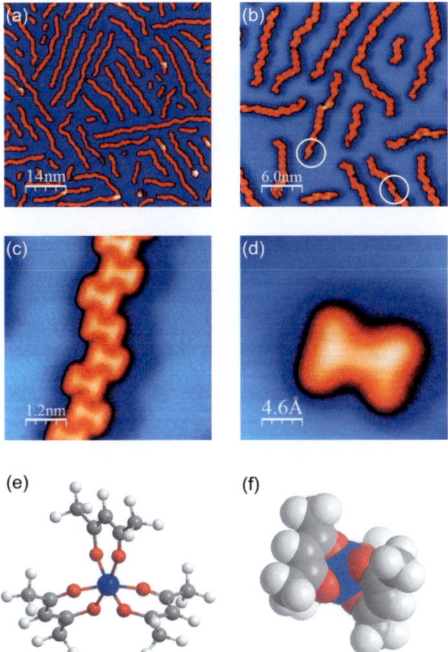

Figure 6.2: Topographical images of Cr(acac)$_3$ on Cu(111). (a) The large scale scan show the alignment in rows along the symmetry axises of the surface (0.8 V, 1.7 nA). (b) Zooming in reveals the individual molecules. They are well ordered over a long range of the chains with only a few defects (100 mV, 1.5 nA). (c) The individual molecules appear as H-shaped chain links that are connected (1.0 V, 1.5 nA). (d) Rarely individual Cr(acac)$_3$ can be observed (100 mV, 1.3 nA). (e) Schematic image of the Cr(acac)$_3$ in vacuum (blue = Cr, red = O, gray = C, white = H). (f) Sketch displaying how the molecule lies on the surface.

molecules form chains consisting of single molecules — each is $\sim 0.9\,$nm long — up to chain length of 50 nm. Thus, chains consisting of $1 - 60$ molecules are formed. In Figure 6.1 the distribution of the chain lengths obtained from several topographies is shown. There is a maximum for short chains corresponding to chains consisting of $2-8$ molecules. The chains are very regular over a long range only showing few branchings or defects. Some molecules seam to be adsorbed on two of the acac ligands. For these molecules only one ligand pointing out of the surface can be seen in the STM image, indicated by the white circles in Figure 6.2 (b). However, a few times it was also possible to observe individual molecules (see Fig. 6.2 (d)) where the H like shape of the individual molecule can be seen clearly.

6.2 Electronic properties

The adsorption studies on Cr(acac)$_3$ revealed that the molecules grow as chains rather than as individual molecules. This could be even an advantage for a controlled deposition of nanostructured magnetic bits on a surface. Thus, the Cr(acac)$_3$ chains are used to examine the electronic and magnetic properties of the molecule — especially of the Cr ion. For this purpose, elastic tunneling spectra were obtained. Since the metallic center and the ligands were physically completely different, the results of the two parts had to be considered separately. Regarding the spectra shown in Figure 6.3 there were three main features that could be observed: the HOMO and LUMO, plus a zero bias resonance, which is only visible for the spectra taken on the Cr ion.

HOMO and LUMO of Cr(acac)$_3$

At first, the focus is on the HOMO and LUMO, which are displayed at $-1.2\,$eV and $1.0\,$eV, respectively. Spatial resolved dI/dV maps at these energies were obtained to visualize the molecular orbitals. In Fig. 6.4 the orbitals are shown for a part of a longer chain (a $-$ c) and a molecular dimer (d $-$ f). For both, the LUMO is mainly concentrated at the links between the molecules and at the end of the chains (see Fig. 6.4 (c) and (f)). The HOMO is always located only on one side of the molecule and is slightly distorted. It is also less pronounced than the LUMO, what can be seen in the dI/dV curve, too.

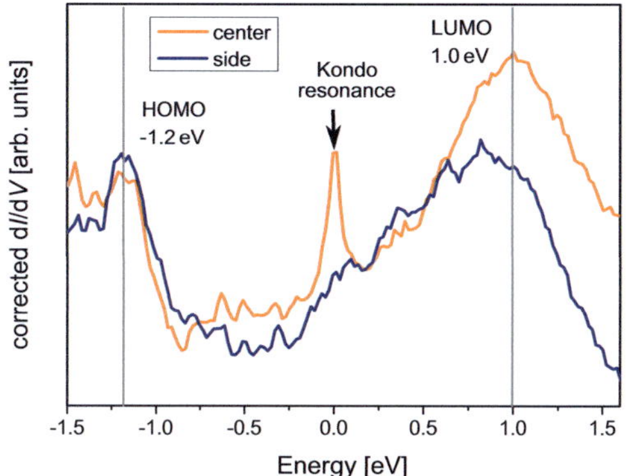

Figure 6.3: dI/dV curve measured on top of the center and the sidegroup of a $Cr(acac)_3$ molecule at $T = 4.3\,K$. The spectra is corrected after the method of Ukraintsev [70]. Three main features can be seen: the HOMO at $-1.2\,meV$, the Kondo resonance $0\,meV$, and the LUMO at $1.0\,meV$

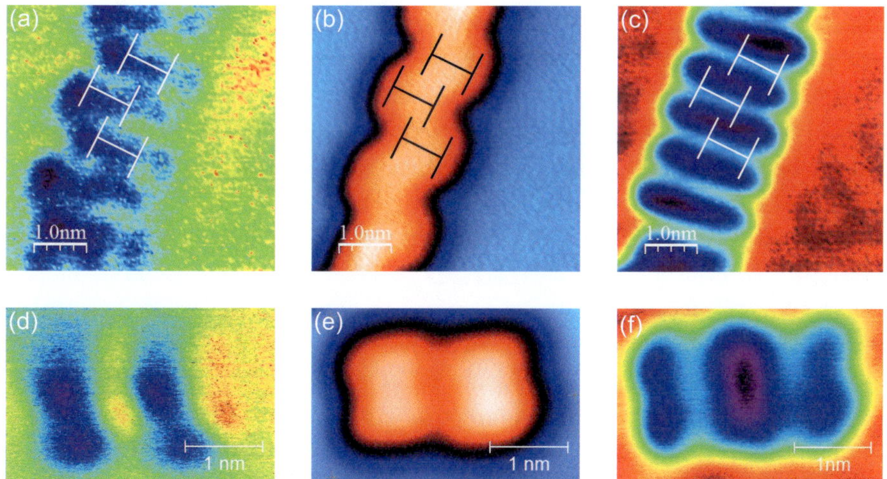

Figure 6.4: (a) HOMO ($-1.2\,eV$), (b) topography, and (c) LUMO ($1.2\,eV$) of a $Cr(acac)_3$ chain. The H shaped lines indicate the position of the molecules ($T = 4.3\,K$). (d) HOMO ($-1.2\,eV$), (e) topography, and (f) LUMO ($1.11\,eV$) of a $Cr(acac)_3$ dimer ($T = 4.3\,K$).

The Kondo resonance

The third and most obvious observation in Figure 6.3 is the zero bias peak. This peak is mainly present for dI/dV curves measured on the center of the molecule and thus must be connected to the Cr ion. Since the zero field ground state of the Cr^{3+} ion is a degenerate spin-$1/2$ doublet [25], this is an indication for a Kondo effect. Thus, a stable spin is not expected for the $Cr(acac)_3$ molecules on a $Cu(111)$ surface. In order to further examine the peak, close-up spectra around the Fermi edge ($\pm 50\,mV$) were measured (see Fig. 6.5 (a)). The peak clearly shows a Fano like shape, typical for the Kondo resonance [132, 133]. Thus the Fano function [2.7] was used to fit the peak [134, 135]. The peak position was fitted to be at $\varepsilon_0 = 0\,meV$. Further, the Kondo temperature could be extracted from the fit to be

$$T_K = k_B (13.0 \pm 0.4)\,meV = (151 \pm 5)\,K. \qquad [6.1]$$

This is in the same energy range as for FePc or CuPc on $Au(111)$ [66, 136]. Further, the fitted $q = 0.19$ of the system is in the region of intermediate coupling for systems showing Kondo effect. Nevertheless, the strong coupling of the Cr ion with the surface resulting in the Kondo effect reveals that the $Cr(acac)_3$ molecules are not a good choice for using them as stable bits. Since the Kondo effect quenches the spin of $Cr(acac)_3$ on $Cu(111)$ it is impossible to build a stable bit out of them. Spacial resolved measurements of the differential conductance were used to localize the Kondo resonance inside the molecular chain. Fig. 6.5 (b) and (c) show the topography and the corresponding dI/dV map at the Fermi edge. The bright areas can be clearly located at the Cr center of the molecule. This is simultaneously an additional proof of the postulated adsorption geometry and of the Kondo effect, which should be located at the central metal ion with spin $1/2$. Additionally to the Kondo resonances on the Cr ions, there are peaks visible in the middle of the kinks in the Kondo chain, indicated in Figure 6.5 (c) by the orange arrows. This occurs when the three surrounding Kondo resonances are at the same distance of around 1.3nm. The green arrows indicating a third group of maxima in the dI/dV map at twice the distance.

A similar effect was observed by Manoharan and co-workers for the mirroring of a Kondo resonance inside a quantum coral formed [26]. A quantum coral is an ellipse built of individual adatoms, *e. g.* Co atoms, on the surface [137]. When for example an additional Co atom is placed in one of the foci of the ellipse, the Kondo resonance is also visible in the other one [26]. Since there is no second atom at this point, there must be a mechanism that is responsible for the transmission of the Kondo states in the surface.

At the (111) surfaces of noble metals *i. e.* Ag(111), Au(111), and in particular Cu(111), Shockley surface states are formed due to the distinct shape of the Fermi surfaces [138]. These surface states can be directly seen as standing waves in STM images at step edges or around nanostructures [139]. Inside the quantum coral the adatoms forming a confined area in which a standing wave pattern is formed [140]. Further, for Kondo resonances on the (111) surfaces of noble metals, the surface state plays an important role [141]. The surface states simultaneously interact with the Kondo impurity and form the standing wave pattern inside the quantum coral. For surface states whose wave functions have constructive interference maxima at

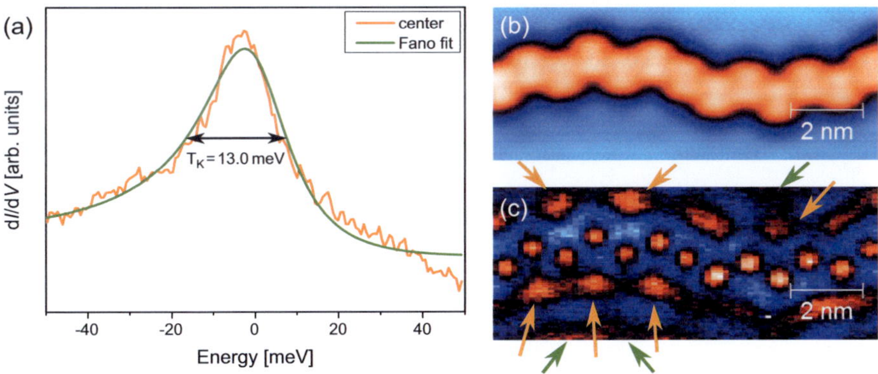

Figure 6.5: (a) Close-up dI/dV scan at the Fermi edge to visualize the Fano shaped peak and to fit a Fano resonance ($T = 4.3\,\mathrm{K}$). (b) Topography and (c) dI/dV map for a Cr(acac)$_3$ chain at the Fermi edge. The Kondo peaks are clearly localized on the central sites. The orange arrows indicate additional peaks which are located close to the molecular chains on the surface. The green arrows indicate second peaks surrounding the molecular chain. These areas are connected to the 2nd maxima of the interference.

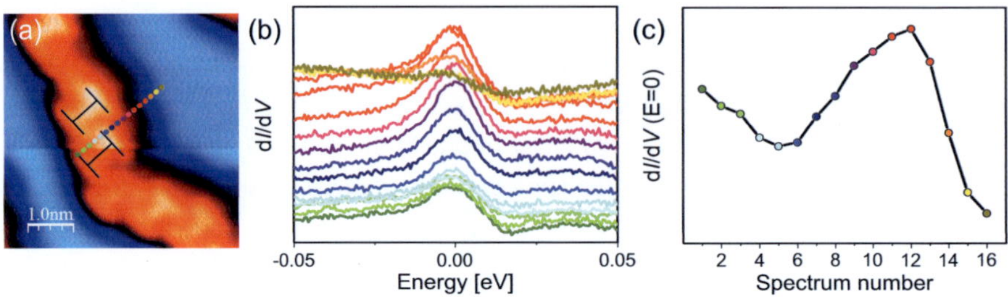

Figure 6.6: (a) Topographic image displaying the point were the spectra are recorded along a line across the molecule away from the Kondo impurity. (b) dI/dV spectra for the different points (each line is shifted slightly to reduce the overlap). (c) Strength of the Kondo peaks of the different spectra. The strongest Kondo peak can be seen at the edge of the molecule, where the resonance is displayed in the Kondo maps (see Fig. 6.5).

both foci the Kondo resonance is mirrored from one to the other [142]. It was theoretically proven that this is really a surface effect and not a effect due to the STM geometry [143].

Similar interference effects of the surface states as in the quantum coral are responsible for the additional peaks that are visible close to the molecular $Cr(acac)_3$ chain on $Cu(111)$. A detailed study of the distance between the central Kondo peak and the additional peaks showed an average distance of

$$d = (12.6 \pm 1.2)\,\text{Å}. \tag{6.2}$$

For interference effects the distance between the maxima is connected to the wavelength

$$\frac{\lambda}{2} = d \tag{6.3}$$

and thus to the wave vector

$$k = \frac{2\pi}{\lambda} = \frac{\pi}{d} = (0.25 \pm 0.02)\,\text{Å}^{-1}. \tag{6.4}$$

This is in good agreement with Fermi wave vector for the band structure of the $Cu(111)$ surface state ($k_F = 0.21\,\text{Å}^{-1}$ [144]).

For a closer study of this effect, different spectra along a line on the molecule away from the dominating center were recorded (see Fig. 6.6 (b)). The intensity of the peak on the surface away from the molecule vanishes. On top of the molecule the peak is visible. It is even sharper as in the areas where constructive interference occurs for several Kondo impurities (see Fig. 6.6 (c)).

7 Conclusions and outlook

In current electronic applications, devices based on single molecules still play a subordinate role. However, the fundamental research has progressed faster in the last years on the way to future applications. The concepts showed to have a high potential to contribute to the process of improving current devices. The STM showed to be an important tool for the research on molecular systems. It combines the possibilities to study growth, transport, as well as magnetic and electronic properties of various systems such as molecular films, molecular nanostructures, or even individual molecules. In the present work, a variety of STM techniques were used to extend the current knowledge. This work focused on three main parts: non-magnetic transport across different phthalocyanine molecules, spin dependent transport across H_2Pc and finally electronic and magnetic studies on the $Cr(acac)_3$ molecules, carrying spin $1/2$.

The first goal was to establish the transport measurements with the STM by taking current-distance curves. This method was already used by different groups to study transport across single molecules [9,16,17,19,80]. The molecules of choice were the phthalocyanine molecules. Due to their flat geometry, they showed a clear indication of the contact formation in the form of a discontinuity in the current-distance curves as observed for similar molecules [18,20]. To separate the tunneling and the ballistic contribution to the transport, the molecular conductance was calculated based on the model of Haiss and co-workers [18]. The individual conductances obtained from numerous measurements on the different systems showed a narrow distribution, especially when compared to the ones obtained with break junction measurements [145].

With this method it was possible to study different influences on the transport properties of phthalocyanine molecules. The average conductances found for the phthalocyanines revealed to be in the same range than C_{60} [19] and a magnitude higher compared for example to the benzene chains [18]. By putting different metal ions in the central cavities, the conductance (on Cu) is reduced from $0.176 G_0$ for the metal free H_2Pc to $0.063\,G_0$ for CoPc and to $0.123\,G_0$ for MnPc. The main difference between the H_2Pc and CoPc was examined by measuring elastic tunneling spectra. The Co ion has a strong influence on the HOMO of the molecule which is shifted further away from the Fermi edge. This reduces the overlap of the orbital with the Fermi level and thus its contribution to the transport. The comparison of theoretical calculations for the molecular orbitals in the gas phase [122, 123] suggests that the HOMO-LUMO gap of MnPc is also smaller than the one of CoPc. This explains the better electron transport in the case of MnPc compared to CoPc. Overall, it can be concluded that the metal ion changes the orbital structure of the molecule and reduces the transmission over the molecule from one interface to the other. The strength of the influence depends on the different metal ions that are used to form the metal-organic compound.

The second main influence on the transport across a molecule sandwiched between two electrodes is the coupling between the different parts of the junction. This is primarily determined by the hybridization of the molecule with the electrodes. The measurements on two different substrates showed this strong influence. The hybridization on Co is much stronger than on Cu: for both H_2Pc and CoPc the bonding to the surface is enhanced. The stronger bonding causes a shift of the LUMO to the Fermi edge. This results in an strong increase of the conductance, for example for H_2Pc from $0.176\,G_0$ (Cu) to $0.296\,G_0$ (Co). Theoretical calculations allowed to connect the shift to an electron transfer from the nitrogen atoms of the molecule to the surface.

For both effects the STM has proven its capabilities to address properties beyond the transport measurement by obtaining elastic tunneling spectra. An additional feature of the STM is inelastic tunneling spectroscopy that could contribute to the studies of molecular transport. The measurements not only revealed a vibron excitation of the molecule. By using spatially resolved ITS, the excitation could be localized on the sidegroups of the molecule. This could be used to explain the formation of the contact, which is consistent with the local minima of the theoretical optimization of the molecular geometry in contact.

Besides the fundamental electron transport in modern applications, magnetoresistance effects often play an essential role. Especially for data storage devices the size of the GMR components is of fundamental importance. In this work the first measurement on a GMR junction consisting of a single molecule is reported. Based on the non-magnetic measurements spin polarized transport measurements were carried out on H_2Pc molecules. By combining a spin polarized STM with contact measurements the dependence of the transport on the relative alignment of the spin polarization of tip and sample was studied. The pioneering measurement of the GMR of an individual molecule in the ballistic transport regime could be successfully performed. A remarkably large GMR ratio (61 %) — one magnitude larger than the TMR of the Co/vacuum/Co junction — was measured for a ballistic Co/H_2Pc/Co junction. This is not only much higher than the TMR for the same electrodes, it also exceeds values of GMR read heads in modern hard disks ($\sim 10 - 20\%$ [146, 147]). For possible future applications, it is of fundamental importance to provide devices of a sufficiently small size and concurrently a high GMR ratio. The phthalocyanine molecules which are in the size of a few nanometers ($\sim 2\,nm$) meet both demands.

Since optimization for future applications needs a detailed understanding of the physical processes underlying the effect, theoretical calculations were performed. It was possible to connect the strong influence of the non-magnetic H_2Pc molecule on the magnetotransport with the distinct interaction of the molecule with the surface. The hybridization between the LUMO of the molecule and the states of the Co substrate is much stronger for the minority channel. Thus, the state is broader in energy and has a stronger overlap with the Fermi level. This effect is illustrated best with the real space images of the important states. The strongest contribution results from the minority channel. For parallel aligned electrodes it forms a state that is completely extended from the tip to the Co islands. The theoretical value (65 %) shows a good consistency with the experimental result (61 %).

In order to optimize the properties, a second molecular GMR junction was examined. For this measurement an asymmetric geometry was employed. For one electrode antiferromagnetic Mn was used while the second electrode consisted of ferromagnetic Fe. While it was not possible to increase the GMR ratio, another interesting effect was discovered. Due to the asymmetry of the junction a negative GMR was observed ($\sim -54\%$). This can be explained by a distinct hybridization with the two electrodes which can cause a change of sign for the GMR when the different spin channels hybridize stronger for the different materials. However, for a complete understanding of the process theoretical calculations would be needed. Another possible application of single molecules in data storage is to build individual bits out of them. The goal was to examine individual molecules on the surface that carry a spin. These experiments where performed using $Cr(acac)_3$ molecules. Instead of having a stable spin, the molecules revealed a Kondo effect due to strong coupling of the spin with the surface. The Kondo effect was not only located at the Cr ions of the molecules. Due to mediation of the surface states the Kondo resonances occurred at different interference maxima periodically surrounding the molecule in intervals of ~ 1.3 nm and multiples of it. This can be connected to the Fermi wave vectors of the surface state. This is a similar effect to that seen in quantum mirages [26].

Overall, the STM has proven its various possibilities to contribute not only in the transport studies but also in complementary studies. It was possible to measure the magnetoresistance of a junction consisting of an individual molecule in the ballistic regime. This could have a big impact on the way of increasing data storage devices and other applications of GMR sensors. Further experiments have to be performed to optimize geometries for such molecular junctions: increase GMR ratios, optimize preparation, design adequate contact geometries. The $Cr(acac)_3$ molecules did not show a stable spin on the Cu(111) surface. To stabilize the spin, insulating surfaces could prohibit the Kondo effect which quenches the spin. However, the Kondo chains showed nice effects that could be used for further studies of the surface state mediated Kondo resonances.

What remains are the engineering tasks to design devices for application in different sensors. But the GMR sensors are not only limited to applications in hard disks. They have been successfully implemented in different other fields, for example in automotive applications [32, 33] or in biosensing [34]. Especially for the latter the fact that the GMR junction is built of an individual molecule could be an interesting new approach.

Bibliography

[1] BAIBICH, M. N., BROTO, J. M., FERT, A., VAN DAU, F. N., PETROFF, F., ETIENNE, P., CREUZET, G., FRIEDERICH, A., and CHAZELAS, J. Giant Magnetoresistance of (001)Fe/(001)Cr Magnetic Superlattices. *Phys. Rev. Lett.* **61**, 2472 (1988).

[2] BINASCH, G., GRÜNBERG, P., SAURENBACH, F., and ZINN, W. Enhanced magnetoresistance in layered magnetic structures with antiferromagnetic interlayer exchange. *Phys. Rev. B* **39**, 4828 (1989).

[3] WD Mobile Hard Drives with Advanced Format Technology Offer Highest Capacity For Mainstream Notebook Computers. URL `http://www.wdc.com/en/company/releases/PressRelease.asp?release=5ba85c8b-0849-4158-8d72-b1be7ee43852`.

[4] YUASA, S., NAGAHAMA, T., FUKUSHIMA, A., SUZUKI, Y., , and ANDO, K. Giant room-temperature magnetoresistance in single-crystal Fe/MgO/Fe magnetic tunnel junctions. *Nature Mat.* **3**, 868 (2004).

[5] PARKIN, S. S. P., KAISER, C., PANCHULA, A., RICE, P. M., HUGHES, B., SAMANT, M., and YANG, S.-H. Giant tunnelling magnetoresistance at room temperature with MgO (100) tunnel barriers. *Nature Mat.* **3**, 862 (2004).

[6] Hitachi achieves Nanotechnology Milestone for Quadrupling Terabyte Hard Drive. URL `http://www.hitachi.com/New/cnews/071015a.html`.

[7] AVIRAM, A. and RATNER, M. A. Molecular rectifiers. *Chem. Phys. Lett.* **29**, 277 (1974).

[8] REED, M. A., ZHOU, C., MULLER, C. J., BURGIN, T. P., and TOUR, J. M. Conductance of a Molecular Junction. *Science* **278**, 252 (1997).

[9] JOACHIM, C., GIMZEWSKI, J. K., SCHLITTLER, R. R., and CHAVY, C. Electronic Transparence of a Single C_{60} Molecule. *Phys. Rev. Lett.* **74**, 2102 (1995).

[10] BUMM, L. A., ARNOLD, J. J., CYGAN, M. T., DUNBAR, T. D., BURGIN, T. P., JONES, I., L., ALLARA, D. L., TOUR, J. M., and WEISS, P. S. Are Single Molecular Wires Conducting? *Science* **271**, 1705 (1996).

[11] NG, M.-K., LEE, D.-C., and YU, L. Molecular Diodes Based on Conjugated Diblock Co-oligomers. *Journal of the American Chemical Society* **124**, 11862 (2002).

[12] ELBING, M., OCHS, R., KOENTOPP, M., FISCHER, M., VON HÄNISCH, C., WEIGEND, F., EVERS, F., WEBER, H. B., and MAYOR, M. A single molecule diode. *Proc. Natl. Acad. Sci. USA* **102**, 8815 (2005).

[13] CHEN, J., REED, M. A., RAWLETT, A. M., and TOUR, J. M. Large On-Off Ratios and Negative Differential Resistance in a Molecular Electronic Device. *Science* **286**, 1550 (1999).

[14] DONHAUSER, Z. J., MANTOOTH, B. A., KELLY, K. F., BUMM, L. A., MONNELL, J. D., STAPLETON, J. J., PRICE, J., D. W., RAWLETT, A. M., ALLARA, D. L., TOUR, J. M., and WEISS, P. S. Conductance Switching in Single Molecules Through Conformational Changes. *Science* **292**, 2303 (2001).

[15] WIESENDANGER, R., GÜNTHERODT, H.-J., GÜNTHERODT, G., GAMBINO, R. J., and RUF, R. Observation of vacuum tunneling of spin-polarized electrons with the scanning tunneling microscope. *Phys. Rev. Lett.* **65**, 247 (1990).

[16] TANG, H., CUBERES, M. T., JOACHIM, C., and GIMZEWSKI, J. K. Fundamental considerations in the manipulation of a single C_{60} molecule on a surface with an STM. *Surface Science* **386**, 115 (1997).

[17] PRADHAN, N. A., LIU, N., and HO, W. Vibronic Spectroscopy of Single C_{60} Molecules and Monolayers with the STM. *The Journal of Physical Chemistry B* **109**, 8513 (2005).

[18] HAISS, W., WANG, C., GRACE, I., BATSANOV, A. S., SCHIFFRIN, D. J., HIGGINS, S. J., BRYCE, M. R., LAMBERT, C. J., and NICHOLS, R. J. Precision control of single-molecule electrical junctions. *Nature Mat.* **5**, 995 (2006).

[19] NÉEL, N., KRÖGER, J., LIMOT, L., FREDERIKSEN, T., BRANDBYGE, M., and BERNDT, R. Controlled Contact to a C_{60} Molecule. *Phys. Rev. Lett.* **98**, 065502 (2007).

[20] TEMIROV, R., LASSISE, A., ANDERS, F. B., and TAUTZ, F. S. Kondo effect by controlled cleavage of a single-molecule contact. *Nanotechnology* **19**, 065401 (2008).

[21] STIPE, B. C., REZAEI, M. A., and HO, W. Single-Molecule Vibrational Spectroscopy and Microscopy. *Science* **280**, 1732 (1998).

[22] HIRJIBEHEDIN, C. F., LUTZ, C. P., and HEINRICH, A. J. Spin Coupling in Engineered Atomic Structures. *Science* **312**, 1021 (2006).

[23] BALASHOV, T., TAKÁCS, A. F., WULFHEKEL, W., and KIRSCHNER, J. Magnon Excitation with Spin-Polarized Scanning Tunneling Microscopy. *Phys. Rev. Lett.* **97**, 187201 (2006).

[24] LEZNOFF, C. C. and LEVER, A. B. P. *Phthalocyanines: Properties and Applications.* (John Wiley & Sons, 1989).

[25] MILLER, J., SCHAEFLE, N., and SHARP, R. Calculating NMR paramagnetic relaxation enhancements without adjustable parameters: the spin-3/2 complex Cr(III)(AcAc)$_3$. *Magnetic Resonance in Chemistry* **41**, 806 (2003).

[26] MANOHARAN, H. C., LUTZ, C. P., and EIGLER, D. M. Quantum mirages formed by coherent projection of electronic structure. *Nature* **403**, 512 (2000).

[27] MOORE, G. Cramming More Components Onto Integrated Circuits. *Proceedings of the IEEE* **86**, 82 (1998).

[28] WALTER, C. Kryder's law. *Scientific American* **293**, 32 (2005).

[29] CHAPPERT, C., FERT, A., and DAU, F. N. V. The emergence of spin electronics in data storage. *Nat. Mater.* **6**, 813 (2007).

[30] JULLIÈRE, M. Tunneling between ferromagnetic films. *Physics Letters A* **54**, 225 (1975).

[31] FUKUZAWA, H., YUASA, H., HASHIMOTO, S., IWASAKI, H., and TANAKA, Y. Large magnetoresistance ratio of 10% by $Fe_{50}Co_{50}$ layers for current-confined path current-perpendicular-to-plane giant magnetoresistance spin-valve films. *Appl. Phys. Lett.* **87**, 082507 (2005).

[32] KAPSER, K., ZARUBA, S., SLAMA, P., and KATZMAIER, E. Speed Sensors for Automotive Applications Based on Integrated GMR Technology. In *Advanced Microsystems for Automotive Applications 2008* (edited by J. Valldorf and W. Gessner), VDI-Buch, 211–227 (Springer Berlin Heidelberg, 2008).

[33] DRESSEN, J., BÜRGLER, D. E., and GRÜNBERG, P. Magneto-electronics. In *Technology Guide* (edited by H.-J. Bullinger), 88–91 (Springer Berlin Heidelberg, 2009).

[34] SANDHU, A. Biosensing: New probes offer much faster results. *Nat Nano* **2**, 746 (2007).

[35] DEDIU, V., MURGIA, M., MATACOTTA, F. C., TALIANI, C., and BARBANERA, S. Room temperature spin polarized injection in organic semiconductor. *Solid State Communications* **122**, 181 (2002).

[36] XIONG, Z. H., WU, D., VARDENY, Z. V., and SHI, J. Giant magnetoresistance in organic spin-valves. *Nature* **427**, 821 (2004).

[37] CINCHETTI, M., HEIMER, K., WUSTENBERG, J., ANDREYEV, O., BAUER, M., LACH, S., ZIEGLER, C., GAO, Y., and AESCHLIMANN, M. Determination of spin injection and transport in a ferromagnet/organic semiconductor heterojunction by two-photon photoemission. *Nat Mater* **8**, 115 (2009).

[38] BARRAUD, C., SENEOR, P., MATTANA, R., FUSIL, S., BOUZEHOUANE, K., DERANLOT, C., GRAZIOSI, P., HUESO, L., BERGENTI, I., DEDIU, V., PETROFF, F., and FERT, A. Unravelling the role of the interface for spin injection into organic semiconductors. *Nat Phys* **6**, 615 (2010).

[39] IACOVITA, C., RASTEI, M., HEINRICH, B. W., BRUMME, T., KORTUS, J., LIMOT, L., and BUCHER, J. Visualizating the Spin of Indivitual Cobalt-Phtalocyanine Molecules. *Phys. Rev. Lett.* **101**, 116602 (2009).

[40] GATTESCHI, D., SESSOLI, R., and VILLAIN, J. *Molecular Nanomagnets* (Oxford University Press, 2006).

[41] CAVALLINI, M., FACCHINI, M., ALBONETTI, C., and BISCARINI, F. Single molecule magnets: from thin films to nano-patterns. *Phys. Chem. Chem. Phys.* **10**, 784 (2008).

[42] ROGEZ, G., DONNIO, B., TERAZZI, E., GALLANI, J., KAPPLER, J., BUCHER, J., and DRILLON, M. The Quest for Nanoscale Magnets: The example of [Mn12] Single Molecule Magnets. *Advanced Materials* **21**, 4323 (2009).

[43] BINNIG, G. and ROHRER, H. Scanning tunneling microscopy. *Helv. Phys. Acta* **55**, 726 (1982).

[44] WEISS, P. L'hypothèse du champ moléculaire et la propriété ferromagnétique. *J. de Physique* **6**, 661 (1907).

[45] HEISENBERG, W. Zur Theorie des Ferromagnetismus. *Zeitsch. f. Physik* **38**, 441 (1922).

[46] BLOCH, F. Bemerkung zur Elektronentheorie des Ferromagnetismus und der elektrischen Leitfähigkeit. *Zeitsch. f. Physik* **57**, 545 (1929).

[47] STONER, E. C. Collective Electron Specific Heat and Spin Paramagnetism in Metals. *Proceedings of the Royal Society of London. Series A, Mathematical and Physical Sciences* **154**, 656 (1936).

[48] STONER, E. C. Collective Electron Ferromagnetism. *Proceedings of the Royal Society of London. Series A, Mathematical and Physical Sciences* **165**, 372 (1938).

[49] SLONCZEWSKI, J. C. Conductance and exchange coupling of two ferromagnets separated by a tunneling barrier. *Phys. Rev. B* **39**, 6995 (1989).

[50] MOODERA, J. S., KINDER, L. R., WONG, T. M., and MESERVEY, R. Large Magnetoresistance at Room Temperature in Ferromagnetic Thin Film Tunnel Junctions. *Phys. Rev. Lett.* **74**, 3273 (1995).

[51] MIYAZAKI, T. and TEZUKA, N. Giant magnetic tunneling effect in $Fe/Al_2O_3/Fe$ junction. *Journal of Magnetism and Magnetic Materials* **139**, L231 (1995).

[52] NAGAMINE, Y., MAEHARA, H., TSUNEKAWA, K., DJAYAPRAWIRA, D., WATANABE, N., YUASA, S., and ANDO, K. Ultralow resistance-area product of 0.4 $\Omega(\mu m)^2$ and high magnetoresistance above 50% in CoFeB/MgO/CoFeB magnetic tunnel junction. *Appl. Phys. Lett.* **89**, 162507 (2006).

[53] PARK, J., PASUPATHY, A. N., GOLDSMITH, J. I., CHANG, C., YAISH, Y., PETTA, J. R., RINKOSKI, M., SETHNA, J. P., ABRUNA, H. D., McEUEN, P. L., and RALPH, D. C. Coulomb blockade and the Kondo effect in single-atom transistors. *Nature* **417**, 722 (2002).

[54] DE HAAS, W. J. and VAN DEN BERG, G. J. The Electrical Resistance of Gold and Silver at Low Temperatures. *Physica III* **6**, 440 (1936).

[55] KONDO, J. Resistance Minimum in Dilute Magnetic Alloys. *Prog. theor. phys.* **32**, 37 (1964).

[56] ANDERSON, P. W. Localized Magnetic States in Metals. *Phys. Rev.* **124**, 41 (1961).

[57] GOLDHABER-GORDON, D., SHTRIKMAN, H., MAHALU, D., ABUSCH-MAGDER, D., MEIRAV, U., and KASTNER, M. A. Kondo Effect in a Single-Elektron Transistor. *Nature* **391**, 156 (1998).

[58] CRONENWETT, S. M., OOSTERKAMP, T. H., and KOUWENHOVEN, L. P. A Tunable Konde Effekt in Quantumdots. *Science* **281**, 540 (1998).

[59] KOUWENHOVEN, L. P. and GLAZMAN, L. Revival of the Kondo Effect. *Physics World* **14**, 33 (2001).

[60] NG, T. K. and LEE, P. A. On-Site Coulomb Repulsion and Resonant Tunneling. *Phys. Rev. Lett.* **61**, 1768 (1988).

[61] GLAZMAN, L. and RAIKH, M. E. Resonant Kondo transparency of a barrier with quasilocal impurity states. *JETP Lett.* **47**, 452 (1988).

[62] PAASKE, J., ROSCH, A., WÖLFLE, P., MASON, N., MARCUS, C. M., and NYGÅRD, J. Kondo Physics in Carbon Nanotubes. *Nature* **408**, 342 (2000).

[63] JESPERSEN, T. S., AAGESEN, M., SØRENSEN, C., LINDELOF, P. E., and NYGÅRD, J. Kondo Physics in Tunable Semiconductor Nanowire Quantum Dots. *Phys. Rev. B* **74**, 233304 (2006).

[64] LIANG, W., SHORES, M. P., BOCKRATH, M., LONG, J. R., and PARK, H. Kondo resonance in a single-molecule transistor. *Nature* **417**, 725 (2002).

[65] PARKS, J. J., CHAMPAGNE, A. R., COSTI, T. A., SHUM, W. W., PASUPATHY, A. N., NEUSCAMMAN, E., FLORES-TORRES, S., CORNAGLIA, P. S., ALIGIA, A. A., BAL-SEIRO, C. A., CHAN, G. K.-L., ABRUÑA, H. D., and RALPH, D. C. Mechanical Control of Spin States in Spin-1 Molecules and the Underscreened Kondo Effect **328**, 1370 (2010).

[66] ZHAO, A., LI, Q., CHEN, L., XIANG, H., WANG, W., PAN, S., WANG, B., XIAO, X., YANG, J., HOU, J. G., and ZHU, Q. Controlling the Kondo Effect of an Adsorbed Magnetic Ion Through Its Chemical Bonding. *Science* **309**, 1542 (2005).

[67] ROMEIKE, C., WEGEWIJS, M. R., HOFSTETTER, W., and SCHOELLER, H. Kondo-Transport Spectroscopy of Single Molecule Magnets. *Phys. Rev. Lett.* **97**, 206601 (2006).

[68] BARDEEN, J. Tunnelling from a Many-Particle Point of View. *Phys. Rev. Lett.* **6**, 57 (1961).

[69] TERSOFF, J. and HAMANN, D. R. Theory and Application for the Scanning Tunneling Microscope. *Phys. Rev. Lett.* **50**, 1998 (1983).

[70] UKRAINTSEV, V. A. Data evaluation technique for electron-tunneling spectroscopy. *Phys. Rev. B* **53**, 11176 (1996).

[71] JAKLEVIC, R. C. and LAMBE, J. Molecular Vibration Spectra by Electron Tunneling. *Phys. Rev. Lett.* **17**, 1139 (1966).

[72] KLEIN, J., LÉGER, A., BELIN, M., DÉFOURNEAU, D., and SANGSTER, M. J. L. Inelastic-Electron-Tunneling Spectroscopy of Metal-Insulator-Metal Junctions. *Phys. Rev. B* **7**, 2336 (1973).

[73] WULFHEKEL, W. and KIRSCHNER, J. Spin-polarized scanning tunneling microscopy on ferromagnets. *Applied Physics Letters* **75**, 1944 (1999).

[74] SCHLICKUM, U., WULFHEKEL, W., and KIRSCHNER, J. Spin-polarized scanning tunneling microscope for imaging the in-plane magnetization. *Applied Physics Letters* **83**, 2016 (2003).

[75] STROSCIO, J. A., PIERCE, D. T., DAVIES, A., CELOTTA, R. J., and WEINERT, M. Tunneling Spectroscopy of bcc (001) Surface States. *Phys. Rev. Lett.* **75**, 2960 (1995).

[76] PIETZSCH, O., KUBETZKA, A., BODE, M., and WIESENDANGER, R. Spin-Polarized Scanning Tunneling Spectroscopy of Nanoscale Cobalt Islands on Cu(111). *Phys. Rev. Lett.* **92**, 057202 (2004).

[77] YERSIN, H. *Highly efficient OLEDs with phosphorescent materials* (Wiley-VCH, 2007).

[78] SUN, S. and SARICIFTCI, N. S. *Organic Photovoltaics: Mechanisms, Materials, and Devices: Mechanism, Materials, and Devices* (Marcel Dekker Inc, 2005).

[79] PARK, H., PARK, J., LIM, A. K. L., ANDERSON, E. H., ALIVISATOS, A. P., and MCEUEN, P. L. Nanomechanical oscillations in a single-C_{60} transistor. *Nature* **407**, 57 (2000).

[80] HAISS, W., NICHOLS, R. J., VAN ZALINGE, H., HIGGINS, S. J., BETHELL, D., and SCHIFFRIN, D. J. Measurement of single molecule conductivity using the spontaneous formation of molecular wires. *Phys. Chem. Chem. Phys.* **6**, 4330 (2004).

[81] PARK, H., LIM, A. K. L., ALIVISATOS, A. P., PARK, J., and MCEUEN, P. L. Fabrication of metallic electrodes with nanometer separation by electromigration. *Applied Physics Letters* **75**, 301 (1999).

[82] CUI, X. D., PRIMAK, A., ZARATE, X., TOMFOHR, J., SANKEY, O. F., MOORE, A. L., MOORE, T. A., GUST, D., HARRIS, G., and LINDSAY, S. M. Reproducible Measurement of Single-Molecule Conductivity. *Science* **294**, 571 (2001).

[83] RAWLETT, A. M., HOPSON, T. J., AMLANI, I., ZHANG, R., TRESEK, J., NAGAHARA, L. A., TSUI, R. K., and GORONKIN, H. A molecular electronics toolbox. *Nanotechnology* **14**, 377 (2003).

[84] ZHOU, C., DESHPANDE, M. R., REED, M. A., II, L. J., and TOUR, J. M. Nanoscale metal/self-assembled monolayer/metal heterostructures. *Applied Physics Letters* **71**, 611 (1997).

[85] KUSHMERICK, J. G., HOLT, D. B., YANG, J. C., NACIRI, J., MOORE, M. H., and SHASHIDHAR, R. Metal-Molecule Contacts and Charge Transport across Monomolecular Layers: Measurement and Theory. *Phys. Rev. Lett.* **89**, 086802 (2002).

[86] METZGER, R. M., CHEN, B., HÖPFNER, U., LAKSHMIKANTHAM, M. V., VUILLAUME, D., KAWAI, T., WU, X., TACHIBANA, H., HUGHES, T. V., SAKURAI, H., BALDWIN, J. W., HOSCH, C., CAVA, M. P., BREHMER, L., and ASHWELL, G. J. Unimolecular Electrical Rectification in Hexadecylquinolinium Tricyanoquinodimethanide. *Journal of the American Chemical Society* **119**, 10455 (1997).

[87] PETTA, J. R., SLATER, S. K., and RALPH, D. C. Spin-Dependent Transport in Molecular Tunnel Junctions. *Phys. Rev. Lett.* **93**, 136601 (2004).

[88] DEDIU, V. A., HUESO, L. E., BERGENTI, I., and TALIANI, C. Spin routes in organic semiconductors. *Nat Mater* **8**, 707 (2009).

[89] SZULCZEWSKI, G., TOKUC, H., OGUZ, K., and COEY, J. M. D. Magnetoresistance in magnetic tunnel junctions with an organic barrier and an MgO spin filter. *Applied Physics Letters* **95**, 202506 (2009).

[90] ROCHA, A. R., GARCIA-SUAREZ, V. M., BAILEY, S. W., LAMBERT, C. J., FERRER, J., and SANVITO, S. Towards molecular spintronics. *Nat Mater* **4**, 335 (2005).

[91] SZULCZEWSKI, G., SANVITO, S., and COEY, M. A spin of their own. *Nat Mater* **8**, 693 (2009).

[92] ATODIRESEI, N., BREDE, J., LAZIC, P., CACIUC, V., HOFFMANN, G., WIESENDANGER, R., and BLÜGEL, S. Controlling the Local Spin-Polarization at the Organic-Ferromagnetic Interface. *Arxiv preprint arXiv:1005.5411* (2010).

[93] BALASHOV, T. *Inelastic scanning tunneling spectroscopy : magnetic excitations on the nanoscale*. Ph.D. thesis, Universität Karlsruhe (TH) (2009).

[94] Davisson, C. and Germer, L. H. Diffraction of Electrons by a Crystal of Nickel. *Phys. Rev.* **30**, 705 (1927).

[95] Lander, J. J. Auger Peaks in the Energy Spectra of Secondary Electrons from Various Materials. *Phys. Rev.* **91**, 1382 (1953).

[96] Harris, L. A. Analysis of Materials by Electron-Excited Auger Electrons. *Journal of Applied Physics* **39**, 1419 (1968).

[97] Moon, A. R. and Cowley, J. M. Medium Energy Electron Diffraction. *Journal of Vacuum Science and Technology* **9**, 649 (1972).

[98] Qiu, Z. Q. and Bader, S. D. Surface magneto-optic Kerr effect. *Review of Scientific Instruments* **71**, 1243 (2000).

[99] de la Figuera, J., Prieto, J. E., Ocal, C., and Miranda, R. Surface etching and enhanced diffusion during the early stages of the growth of Co on Cu(111). *Surface Science* **307-309**, 538 (1994).

[100] Vázquez de Parga, A. L., García-Vidal, F. J., and Miranda, R. Detecting Electronic States at Stacking Faults in Magnetic Thin Films by Tunneling Spectroscopy. *Phys. Rev. Lett.* **85**, 4365 (2000).

[101] Diekhöner, L., Schneider, M. A., Baranov, A. N., Stepanyuk, V. S., Bruno, P., and Kern, K. Surface States of Cobalt Nanoislands on Cu(111). *Phys. Rev. Lett.* **90**, 236801 (2003).

[102] Walker, T. G. and Hopster, H. Magnetism of Mn layers on Fe(100). *Phys. Rev. B* **48**, 3563 (1993).

[103] Andrieu, S., Foy, E., Fischer, H., Alnot, M., Chevrier, F., Krill, G., and Piecuch, M. Effect of O contamination on magnetic properties of ultrathin Mn films grown on (001) Fe. *Phys. Rev. B* **58**, 8210 (1998).

[104] Yamada, T. K., Bischoff, M. M. J., Heijnen, G. M. M., Mizoguchi, T., and van Kempen, H. Observation of Spin-Polarized Surface States on Ultrathin bct Mn(001) Films by Spin-Polarized Scanning Tunneling Spectroscopy. *Phys. Rev. Lett.* **90**, 056803 (2003).

[105] SCHLICKUM, U., JANKE-GILMAN, N., WULFHEKEL, W., and KIRSCHNER, J. Step-Induced Frustration of Antiferromagnetic Order in Mn on Fe(001). *Phys. Rev. Lett.* **92**, 107203 (2004).

[106] TULCHINSKY, D. A., UNGURIS, J., and CELOTTA, R. J. Growth and magnetic oscillatory exchange coupling of Mn/Fe(001) and Fe/Mn/Fe(001). *Journal of Magnetism and Magnetic Materials* **212**, 91 (2000).

[107] DELACOTE, G. M., FILLARD, J. P., and MARCO, F. J. Electron injection in thin films of copper phtalocyanine. *Solid State Communications* **2**, 373 (1964).

[108] BRAUN, A. and TCHERNIAC, J. Über die Produkte der Einwirkung von Acetanhydrid auf Phthalamid. *Berichte der deutschen chemischen Gesellschaft* **40**, 2709 (1907).

[109] LÖBBERT, G. *Phthalocyanines* (Wiley-VCH Verlag, 2000).

[110] HAMANN, C., MRWA, A., MÜLLER, M., GÖPEL, W., and RAGER, M. Lead phthalocyanine thin films for NO_2 sensors. *Sensors and Actuators B: Chemical* **4**, 73 (1991).

[111] GHOSH, A. K., MOREL, D. L., FENG, T., SHAW, R. F., and CHARLES A. ROWE, J. Photovoltaic and rectification properties of Al/Mg phthalocyanine/Ag Schottky-barrier cells. *Journal of Applied Physics* **45**, 230 (1974).

[112] DE LA TORRE, G., CLAESSENS, C. G., and TORRES, T. Phthalocyanines: old dyes, new materials. Putting color in nanotechnology. *Chemical Communications* 2000 (2007).

[113] CLAESSENS, C. G., HAHN, U., and TORRES, T. Phthalocyanines: From outstanding electronic properties to emerging applications. *The Chemical Record* **8**, 75 (2008).

[114] GIMZEWSKI, J., STOLL, E., and SCHLITTLER, R. Scanning tunneling microscopy of individual molecules of copper phthalocyanine adsorbed on polycrystalline silver surfaces. *Surface Science* **181**, 267 (1987).

[115] LIPPEL, P. H., WILSON, R. J., MILLER, M. D., WÖLL, C., and CHIANG, S. High-Resolution Imaging of Copper-Phthalocyanine by Scanning-Tunneling Microscopy. *Physical Review Letters* **62**, 171 (1989).

[116] WU, S. W., NAZIN, G. V., CHEN, X., QIU, X. H., and HO, W. Control of Relative Tunneling Rates in Single Molecule Bipolar Electron Transport. *Physical Review Letters* **93**, 236802 (2004).

[117] WANG, J., SHI, Y., CAO, J., and WU, R. Magnetization and magnetic anisotropy of metallophthalocyanine molecules from the first principles calculations. *Applied Physics Letters* **94**, 122502 (2009).

[118] SIDDIQI, M. A., SIDDIQUI, R. A., and ATAKAN, B. Thermal stability, sublimation pressures and diffusion coefficients of some metal acetylacetonates. *Surface and Coatings Technology* **201**, 9055 (2007). Euro CVD 16, 16th European Conference on Chemical Vapor Deposition.

[119] SEMYANNIKOV, P., IGUMENOV, I., TRUBIN, S., CHUSOVA, T., and SEMENOVA, Z. Thermodynamics of chromium acetylacetonate sublimation. *Thermochimica Acta* **432**, 91 (2005).

[120] GRILLO, S. E., TANG, H., COUDRET, C., and GAUTHIER, S. STM observation of the dissociation of a chromium tris-diketonato complex on Cu(100). *Chemical Physics Letters* **355**, 289 (2002).

[121] HEINRICH, B. W., IACOVITA, C., BRUMME, T., CHOI, D.-J., LIMOT, L., RASTEI, M. V., HOFER, W. A., KORTUS, J., and BUCHER, J.-P. Direct Observation of the Tunneling Channels of a Chemisorbed Molecule. *The Journal of Physical Chemistry Letters* **1**, 1517 (2010).

[122] MAROM, N. and KRONIK, L. Density functional theory of transition metal phthalocyanines, II: electronic structure of MnPc and FePc-symmetry and symmetry breaking. *Applied Physics A: Materials Science and Processing* **95**, 165 (2009).

[123] MAROM, N. and KRONIK, L. Density functional theory of transition metal phthalocyanines, I: electronic structure of NiPc and CoPc-self-interaction effects. *Applied Physics A: Materials Science and Processing* **95**, 159 (2009).

[124] Wolf, E. L. *Principles of electron tunneling spectroscopy* (Oxford University Press, 1985).

[125] Pascual, J. I., Jackiw, J. J., Song, Z., Weiss, P. S., Conrad, H., and Rust, H.-P. Adsorbate-Substrate Vibrational Modes of Benzene on Ag(110) Resolved with Scanning Tunneling Spectroscopy. *Phys. Rev. Lett.* **86**, 1050 (2001).

[126] Santos, T., Lee, J., Migdal, P., Lekshmi, I., Satpati, B., and Moodera, J. Room temperature tunnel magnetoresistance and spin polarized tunneling studies with organic semiconductor barrier. *Phys. Rev. Lett.* **98**, 016601 (2007).

[127] Pasupathy, A. N., Bialczak, R. C., Martinek, J., Grose, J. E., Donev, L. A. K., McEuen, P. L., and Ralph, D. C. The Kondo Effect in the Presence of Ferromagnetism. *Science* **306**, 86 (2004).

[128] Oka, H., Ignatiev, P. A., Wedekind, S., Rodary, G., Niebergall, L., Stepanyuk, V. S., Sander, D., and Kirschner, J. Spin-dependent quantum interference within a single magnetic nanostructure. *Science* **327**, 843 (2010).

[129] Arnold, A., Weigend, F., and Evers, F. Quantum chemistry calculations for molecules coupled to reservoirs: Formalism, implementation and application to benzene-dithiol. *J. Chem. Phys.* **126**, 174101 (2007).

[130] Rosa, A., Ricciardi, G., Baerends, E. J., and van Gisbergen, S. J. A. The Optical Spectra of NiP, NiPz, NiTBP, and NiPc: Electronic Effects of Mesotetraaza Substitution and Tetrabenzo Annulation. *The Journal of Physical Chemistry A* **105**, 3311 (2001).

[131] Krusin-Elbaum, L., Shibauchi, T., Argyle, B., Gignac, L., and Weller, D. Stable ultrahigh-density magneto-optical recordings using introduced linear defects. *Nature* **410**, 444 (2001).

[132] Madhavan, V., Chen, W., Jamneala, T., Crommie, M. F., and Wingreen, N. S. Tunneling into a Single Magnetic Atom: Spectroscopic Evidence of the Kondo Resonance. *Science* **280**, 567 (1998).

[133] Újsághy, O., Kroha, J., Szunyogh, L., and Zawadowski, A. Theory of the Fano Resonance in the STM Tunneling Density of States due to a Single Kondo Impurity. *Phys. Rev. Lett.* **85**, 2557 (2000).

[134] Fano, U. Effects of Configuration Interaction on Intensities and Phase Shifts. *Phys. Rev.* **124**, 1866 (1961).

[135] Katoh, K., Yoshida, Y., Yamashita, M., Miyasaka, H., Breedlove, B. K., Kajiwara, T., Takaishi, S., Ishikawa, N., Isshiki, H., Zhang, Y. F., Komeda, T., Yamagishi, M., and Takeya, J. Direct Observation of Lanthanide(III)-Phthalocyanine Molecules on Au(111) by Using Scanning Tunneling Microscopy and Scanning Tunneling Spectroscopy and Thin-Film Field-Effect Transistor Properties of Tb(III)- and Dy(III)-Phthalocyanine Molecules. *Journal of the American Chemical Society* **131**, 9967 (2009).

[136] Gao, L., Ji, W., Hu, Y. B., Cheng, Z. H., Deng, Z. T., Liu, Q., Jiang, N., Lin, X., Guo, W., Du, S. X., Hofer, W. A., Xie, X. C., and Gao, H.-J. Site-Specific Kondo Effect at Ambient Temperatures in Iron-Based Molecules. *Phys. Rev. Lett.* **99**, 106402 (2007).

[137] Crommie, M. F., Lutz, C. P., and Eigler, D. M. Confinement of Electrons to Quantum Corrals on a Metal Surface. *Science* **262**, 218 (1993).

[138] Kevan, S. D. and Gaylord, R. H. High-resolution photoemission study of the electronic structure of the noble-metal (111) surfaces. *Phys. Rev. B* **36**, 5809 (1987).

[139] Hasegawa, Y. and Avouris, P. Direct observation of standing wave formation at surface steps using scanning tunneling spectroscopy. *Phys. Rev. Lett.* **71**, 1071 (1993).

[140] Heller, E. J., Crommie, M. F., Lutz, C. P., and Eigler, D. M. Scattering and absorption of surface electron waves in quantum corrals. *Nature* **369**, 464 (1994).

[141] Merino, J. and Gunnarsson, O. Role of Surface States in Scanning Tunneling Spectroscopy of (111) Metal Surfaces with Kondo Adsorbates. *Phys. Rev. Lett.* **93**, 156601 (2004).

[142] Agam, O. and Schiller, A. Projecting the Kondo Effect: Theory of the Quantum Mirage. *Phys. Rev. Lett.* **86**, 484 (2001).

[143] Fiete, G. A., Hersch, J. S., Heller, E. J., Manoharan, H. C., Lutz, C. P., and Eigler, D. M. Scattering Theory of Kondo Mirages and Observation of Single Kondo Atom Phase Shift. *Phys. Rev. Lett.* **86**, 2392 (2001).

[144] Crommie, M. F., Lutz, C. P., and Eigler, D. M. Imaging standing waves in a two-dimensional electron gas. *Nature* **363**, 524 (1993).

[145] Smit, R. H. M., Noat, Y., Untiedt, C., Lang, N. D., van Hemert, M. C., and van Ruitenbeek, J. M. Measurement of the conductance of a hydrogen molecule. *Nature* **419**, 906 (2002).

[146] Freitas, P., Ferreira, H., Ferreira, R., Cardoso, S., Dijken, S., and Gregg, J. Nanostructures for Spin Electronics. In *Advanced Magnetic Nanostructures* (edited by D. Sellmyer and R. Skomski), 403–460 (Springer US, 2006).

[147] Getzlaff, M. Applications. In *Fundamentals of Magnetism*, 293–335 (Springer Berlin Heidelberg, 2008).

[148] Horcas, I., Fernández, R., Gómez-Rodríguez, J. M., Colchero, J., Gómez-Herrero, J., and Baro, A. M. WSXM: A software for scanning probe microscopy and a tool for nanotechnology. *Review of Scientific Instruments* **78**, 013705 (2007).

[149] Bai, C. *Scanning Tunneling Microscopy and Its Applications* (Springer, 1995).

[150] Mayer, E., Hug, H. J., and Bennewitz, R. *Scanning Probe Microscopy* (Springer, 2004).

[151] Stöhr, J. and Siegmann, H. C. *Magnetism* (Springer, 2006).

[152] Ibach, H. *Physics of surfaces and Interfaces* (Springer, 2006).

Acknowledgments

Thankfully, I did not have to spent the time I needed to carry out this work completely on my own. There are several people without whom this work would not have been possible. This people I want to thank in the following:

- Especially, Prof. Wulf Wulfhekel not only for the possibility to do this work but also for his close contact to the student, always open for questions and to help with different problems. Thus, there was always light at the end of the tunnel.

- Prof. Ferdinand Evers for being the second referee of the work and, together with Alexej Bagrets, for the theoretical support. Without theory the experiment always would be only half as impressive.

- Albert, Timofey, and Toyo for the support not only when starting to work with the STM.

- Tobias and Lukas for the many interesting hours on various conferences and the sharing of the obligations of the oldest PhD student of the group.

- Florian, Annika, Yasmin, and Detlef. Sometimes you cost me some nerves but, nevertheless, it was nice to work with you.

- Toshio for everything that had to be done in the middle of the night.

- Michael for the *Nervennahrung* to overcome smaller and bigger problems.

- The whole group including the former members for the enjoyable working atmosphere.

- The mechanical and electrical workshop for prompt and competent support.

- Martin Bowen, Eric Beaurepaire, and Artur Böttcher with his group for the collaborations on the different molecules.

- I also would like to thank Prof. Pietro Gambardella and his group, especially Cornelius, Timofey, and Jerald, for the nice three month in Barcelona, that allowed me to learn a lot of new things not only on the field of physics.

- The Karlsruhe House of Young Scientists for financial support for my stay in Barcelona.

- Everyone who helped me to erase as much mistakes as possible in this thesis.

- Matthias for the help on the physicist's problems with chemistry.

- All my friends for the whole time in Karlsruhe.

- My parents and my brother for all the support.

- Anja for more than I could write here.